Turbulent MIRROR

Turbulent MIRROR

An Illustrated Guide to Chaos Theory and the Science of Wholeness

JOHN BRIGGS *and* F. DAVID PEAT

Illustrations by Cindy Tavernise

1817

HARPER & ROW, PUBLISHERS, NEW YORK

GRAND RAPIDS, PHILADELPHIA, ST. LOUIS, SAN FRANCISCO

LONDON, SINGAPORE, SYDNEY, TOKYO

We are indebted to those who have given us permission to reprint excerpts from their copyrighted works: "The Writer," by Richard Wilbur, reprinted by permission of the *New Republic*, © 1971, The New Republic, Inc.; "Connoisseur of Chaos," from *The Collected Poems of Wallace Stevens* by Wallace Stevens, © 1942 by Wallace Stevens and renewed 1970 by Holly Stevens, Alfred A. Knopf, Publisher.

 For information address Harper & Row, Publishers, Inc., 10 East 53rd Street, New York, N.Y. 10022. Published simultaneously in Canada by Fitzhenry & Whiteside Limited, Toronto.

FIRST EDITION

DESIGNER: JOEL AVIROM

LIBRARY OF CONGRESS CATALOG CARD NUMBER 88-45567

ISBN 0-06-016061-6

---

89 90 91 92 93 DT/MPC 10 9 8 7 6 5 4 3 2 1

To *Maureen and Barbara*
who were compelled to endure an uncertain amount of chaos
so this book could be written

ACKNOWLEDGMENTS

The authors want to acknowledge with many thanks the following people for their kind assistance on this book:

Ashvin Chhabra and Roderick V. Jensen of the Mason Laboratory for Applied Physics, Yale University; Benoit Mandelbrot and Dennis Arvey of IBM's Thomas J. Watson Research Center in Yorktown Heights, N.Y.; Ilya Prigogine and his colleagues at the Center for Statistical Mechanics, the University of Texas at Austin; Lynn Margulis and Gail Fleischaker of Boston University; Dan Kalikow and David Brooks of Prime Computer; Peter Senge of the Massachusetts Institute of Technology; Douglas Smith of the Boston Museum of Science; Jim Crutchfield of the University of California at Berkeley; Ron Dekett of the Bridgeport Telegram; Frank McCluskey of Mercy College; Charles Redmond and Mike Gentry of NASA; Roy Fairfield of Union Graduate School; networker Laurence Becker; and most especially our editors at Harper & Row, Jeanne Flagg and Rick Kot.

CONTENTS

The Yellow Emperor said: "When my spirit goes through its door, and my bones return to the root from which they grew, what will remain of me?"

CHUANG TZU

The Creation Hymn of the Rig Veda *asserts that in the beginning there was no air, no heavens, no water, no death and no immortality.* Night *and day did not exist and there was only the breathing of the* One. *Then somehow creation occurred.* No *one knows how this happened, and the* Rig Veda *speculates that possibly even the* One *does not know.*

COMMENTARY ON THE RIG VEDA

Humpty Dumpty *growled out,* ". . . *Why, if* I *ever* did *fall off—which there's no chance of—but* if I *did—*" Here *he pursed up his lips, and looked so solemn and grand that* Alice *could hardly help laughing.* "If I did *fall,*" *he went on,* "the King has promised me . . ."

"To *send all his horses and all his men,*" Alice *interrupted, rather unwisely.* . . .

"Yes, *all his horses and all his men,*" Humpty Dumpty *went on.* "They'd *pick me up again in a minute,* they *would!* . . ."

THROUGH THE LOOKING-GLASS

Schopenhauer . . . points out that when you reach an advanced age and look back over your lifetime, it can seem to have had a consistent order and plan, as though composed by some novelist. Events that when they occurred had seemed accidental and of little moment turn out to have been indispensable factors in the composition of a consistent plot. So who composed that plot? Schopenhauer suggests that just as your dreams are composed by an aspect of yourself of which your consciousness is unaware, so, too, your whole life is composed by the will within you. And just as people whom you will have met apparently by mere chance became leading agents in the structuring of your life, so, too, will you have served unknowingly as an agent, giving meaning to the lives of others. The whole thing gears together like one big symphony, with everything unconsciously structuring everything else. . . . one great dream of a single dreamer in which all the dream characters dream, too; . . . Everything arises in mutual relation to everything else, so you can't blame anybody for anything. It is even as though there were a single intention behind it all, which always makes some kind of sense, though none of us knows what the sense might be, or has lived the life that he quite intended.

JOSEPH CAMPBELL

Not Chaos-like, together crushed and bruised,
But, as the world harmoniously confused:
Where order in variety we see,
And where, though all things differ, all agree.

ALEXANDER POPE

FOREWORD

An ancient Chinese legend suggests itself as a metaphor for the puzzles of order and chaos.

In early times, the legend goes, the world of mirrors and the world of humans were not separated as they would be later on. In those days specular beings and human beings were quite different from each other in color and form, though they mingled and lived in harmony. In that time it was also possible to come and go through mirrors. However, one night the mirror people invaded the earth without warning and chaos ensued. Indeed, human beings quickly realized that the mirror people *were* chaos. The power of the invaders was great, and it was only through the magic arts of the Yellow Emperor that they were defeated and driven back to their mirrors. To keep them there the emperor cast a spell that compelled the chaotic beings mechanically to repeat the actions and appearances of men.

The emperor's spell was strong but it would not be eternal, the legend says. The story predicts that one day the spell will weaken and the turbulent shapes in our mirrors will begin to stir. At first the difference between the mirror shapes and our familiar shapes will be unnoticeable. But little by little gestures will separate, colors and forms will transmogrify—and suddenly the long-imprisoned world of chaos will come boiling out into our own.

Perhaps it is already here.

A DC-9 taking off in a snowstorm at Denver's airport encounters trouble only a few feet above ground; it stalls and flips over, killing twenty-eight people. Investigators home in on two alternative explanations for the crash, and both involve new discoveries about the effect of chaotic air currents, or turbulence. In one scenario an unruly vortex of air, spun up in the wake of a jet landing on a nearby runway, failed to dissipate; it lingered for several minutes while other air currents nudged it into the path of the DC-9 and formed a fatal clot in the plane's compressors. In the alternative scenario—which the investigators finally decide is the correct one—the culprit is the few grains of ice that passengers reported seeing on their plane's wings after the final deicing. These small seeds built up a turbulence powerful enough to bring the giant jet down.

Far out at sea, another sort of turbulence makes its appearance. Ordinarily eddies grind and splash and dissipate in the familiar chaos of ocean swells. But researchers have learned that sometimes something happens that seems to violate common sense and the laws of science. In the clash of waves, the watery chaos orchestrates itself, synchronizes its disorders, metamorphoses into a single smooth wave able to travel thousands of miles, beneath ships and through storms, without suffering the slightest loss of shape.

Scientists speculate that yet another form of synchronized chaos may have been at work on the infamous "Black Monday" in October 1987 when worldwide stock prices plummeted. They hypothesize that computer-programmed trading, the computer loop arrangement called portfolio insurance, and the instantaneous communication networks linking financial markets around the world created a situation in which relatively minor bad news rapidly became magnified. For one long day the random and independent behaviors of investors meshed together to create a financial calamity.

As in our version of the Yellow Emperor legend, these examples seem to illustrate that order and chaos are dynamically and mysteriously intertwined. During the past few years, the effort to disentangle that interwovenness has plunged scientists into a new view of reality. This view involves startling insights into nature's wholeness and has forced a reexamination of some of science's most basic assumptions.

The world defined by science traditionally has been a world of almost Platonic purity. The equations and theories describing the rotation of the planets, the rise of water in a tube, the trajectory of a baseball, or the structure of the genetic code contain a regularity and order, a clockworklike certainty, that we have come to associate with nature's laws. Scientists have long admitted, of course, that outside the laboratory our world is seldom as Euclidean as it seems in the mirror of those laws that we hold up to nature. Turbulence, irregularity, and unpredictability are everywhere, but it has always seemed fair to assume that this was "noise," a messiness that resulted from the way things in reality crowd into each other. Put another way, chaos has been thought to be the result of a complexity that in theory could be stripped down to its orderly underpinnings.

Now scientists are discovering that this assumption was mistaken.

A nuthatch pecks erratically for insects that are scattered randomly in the bark of a tree; mountains thrust up and erode into jagged spires clawed at by the forces of an essentially unpredictable long-term weather; the irregular surfaces of hearts, intestines, lungs, and brains join the vast mat of other organic structures that cover the planet in ways not describable in Euclidean terms.

"Most biological systems, and many physical ones, are discontinuous, inhomogeneous, and irregular," says Bruce West, a physicist at the University of California, and Ary Goldberger, a professor at the Harvard Medical School, in an article in *American Scientist*. They are among the growing number of scientists articulating a daring new insight: "The variable, complicated structure and behavior of living systems seem as likely to be verging on chaos as converging on some regular pattern."

Chaos, irregularity, unpredictability. Could it be that such things are not mere noise but have laws of their own? This is what some scientists are now learning. More than that, these scientists are showing how the strange laws of chaos lie behind many if not most of the things we consider remarkable about our world: the human heartbeat and human thoughts, clouds, storms, the structure of galaxies, the creation of a poem, the rise and fall of the gypsy moth caterpillar population, the spread of a forest fire, a winding coastline, even the origins and evolution of life itself.

Thus a new breed of scientists has begun constructing a new mirror to hold up to nature: a turbulent mirror.

In the following pages we will see how in the landscape on one side of that mirror these new researchers are studying the ways in which order falls apart into chaos; how on the other side they are finding out how chaos makes order; and how at the mirror's elusive surface—at the nexus between these worlds—they are helping to shift attention from the quantitative features of dynamical systems to their qualitative properties. And on both sides and at the center, these new scientists are crossing the boundaries of scientific disciplines: Mathematicians are studying biological systems, physicists are taking on problems in neurophysiology; neurophysiologists are boning up on mathematics. Often their common tool is the computer. With it the researchers of chaos iterate equations like chemists combining reagents; colors and shapes representing numbers simmer, congeal, and crackle on terminal screens. Such forms, abstract yet immensely vivid, have helped sharpen unexpected intuitions about how complexity changes. Though we tend to think of the computer as crisp and precise, ironically the computer model with its roiling images of feedback and chaos has become a symbol of a leap the new turbulent science is taking—subordinating scientists' traditional concern with prediction, control, and the analysis of parts to a new concern for the way the unpredictable whole of things moves.

In fact, it is by giving substance to the usually vague term *wholeness* that the science of chaos and change is forging a revolution in our perspective. Reporter and science writer James Gleick says in his fascinating book about the discoveries and personalities of many of the scientists who invented "chaos theory" in the 1970s and 1980s, "More and more [of them] felt the futility of studying parts in isolation from the whole. For them, chaos was the end of the reductionist program in science." A fresh understanding of the concepts of wholeness, chaos, and change is at the heart of the revolution. Chaos physicist Joseph Ford calls it "a major shift in the whole philosophy of science and the way man looks at his world."

So it is that in a few short years the old spell separating the world of chaos from the world of order has seemed to weaken if not dissolve, and science has found itself in the midst of an invasion. Or *is* it an invasion? Perhaps it is something more beneficial and creative, a modern resurgence of the ancient sense of harmony between order and chaos.

ORDER TO *Chaos*

Prologue

An Ancient Tension

The emperor of the South Sea was called Shu (Brief), the emperor of the North Sea was called Hu (Sudden), and the emperor of the central region was called Hun-tun (Chaos). Shu and Hu from time to time came together for a meeting in the territory of Hun-tun, and Hun-tun treated them very generously. Shu and Hu discussed how they could repay his kindness. "All men," they said, "have seven openings so they can see, hear, eat, and breathe. But Hun-tun alone doesn't have any. Let's try boring him some!"

Every day they bored another hole, and on the seventh day Hun-tun died.

CHUANG TZU (BURTON WATSON, TR.)

THE FIRST OF ALL THINGS

Ancient peoples believed that the forces of chaos and order were part of an uneasy tension, a harmony of sorts. They thought of chaos as something immense and creative.

In his *Theogony*, Hesiod assured his audience, "First of all things was *chaos*; and next broad-bosomed Earth." Cosmologies from every culture imagined a primordial state where chaos or nothingness pervaded, from which beings and things burst forth. The ancient Egyptians conceived of the early universe as a formless abyss named Nut. Nut gave birth to Ra, the sun. In one Chinese creation story a ray of pure light, yin, jumps out of chaos and builds the sky while the remaining heavy dimness, yang, forms the earth. Yin and yang, the female and male principles, then act to create the 10,000 things (in other words, everything). Significantly, even after they have emerged, the principles of yin and yang are said to retain the qualities of the chaos from which they sprang. Too much yin or yang will bring chaos back.

In the Babylonian creation story chaos was called Tiāmat. She and other early gods embodied the various faces of chaos. For example, there was a god symbolizing the boundless stretches of primordial formlessness, and a god called "the hidden," representing the intangibility and imperceptibility that lurks in chaotic confusion. The Babylonian insight that the formlessness of chaos could in fact have different faces—in other words, a kind of implicit order—would wait thousands of years to be recovered by modern science.

The turbulent science's fresh-found insight about a reciprocity between order and chaos

Figure P.1. In Chinese myth the dragon represents the principle of order, yang; the dragons are shown here emerging from chaos. But are they emerging? Perhaps they're trying to subdue the disorder or are being subdued by it themselves. The painting illustrates the ancient insight that order and chaos are paradoxical: they are at odds, yet are an integral part of each other.

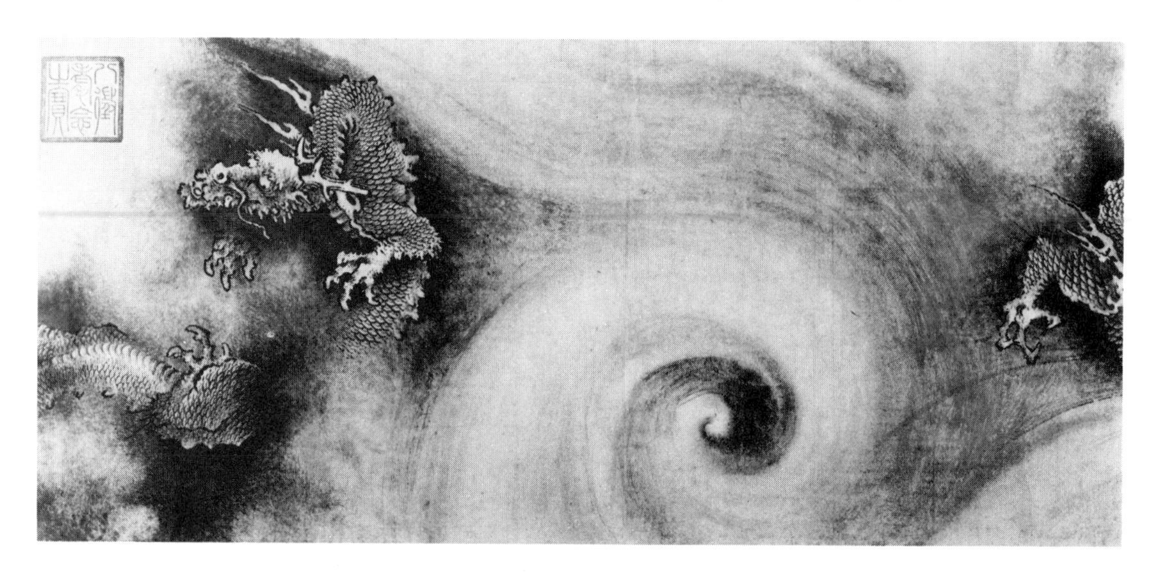

is also a very old idea. Babylonian mythmakers related that as a host of new forms tumbled out of chaos and began to give structure to the universe, Tiāmat became angry. She realized her wonderfully unkempt realm was shrinking. To retrieve her tumultuous territory, she plotted the elimination of all the order she had given birth to. Tiāmat's formless monsters set about terrorizing everything and were successful until Marduk, who was her descendant, defeated her, creating a new kind of order.

The mythic idea that cosmic creativity depends on a reciprocity between order and disorder even survives into the monotheistic cosmologies like Christianity.

Psalm 74:13–14 relates that God (who is order) is compelled to "break the heads of the dragons on the waters" and "crush the heads of Leviathan." One commentator points out that this is the vestige "of a notion of creation which emphasizes the struggle of the deity against the powers of chaos." The Biblical universe starts "without form, and void" until God creates, or orders, it. However, the struggle with disorder is not just a one-time event. The deluge, Satan, the tormentors of Christ, all are manifestations of the hydralike chaos which continues to raise its head. At Christ's crucifixion "the earth shook and rocks were split; and the tombs opened," as disorder threatened again to overtake creation. But perhaps these rumblings of chaos were primarily meant to signal that a new order was on its way. Or, perhaps God's continuing struggle with chaos is really an internal one since, from some perspectives, the Christian creator *himself* is chaos as much as he is order. God is the whirlwind, the fiery destruction, the bringer of plagues and floods. Apparently to be a creator requires operating in a shadowy boundary line between order and chaos. Many cultures have shared this vision. The shape that emerges out of the borderland is Dionysos, the god of random frenzy that underlies the routines of culture; it is the Indian creator god Shiva, who lives in horrible places such as battlefields and crossroads; it is the monsters of sin and death.

While in ancient times the mirror-worlds of chaos and human order lived in an uneasy alliance, science changed all that. With the advent of science—more specifically reductionist science—a spell as powerful as the Yellow Emperor's was cast and a centuries-long suppression of the mirror-world of chaos began.

FORGETTING CHAOS, OR THE MEETING AT HUN-TUN'S

The psychologist, anthropologist, and critic René Girard has observed that we humans have a great need to interpret the disorder in myths from the point of view of order. "Even the word 'dis-order' suggests the precedence and preeminence of order," he says. "We are always improving on mythology in the sense that we suppress its disorder more and more."

One of the ways the early Greek philosophers "improved" on the mythical idea of disorder was by injecting it with a scientific attitude. Thales, Anaximander, and Anaxagoras proposed that a specific substance or energy —water or air—had been in chaotic flux and from that substance the various forms in the universe had congealed. Eventually, so these protoscientists thought, order would dissolve and return to the cosmic flux and then a new universe would appear. This was the old mythic idea made abstract by a clinical attitude.

Aristotle took the scientific approach a step further—and distanced himself even further from chaos. He speculated that order is pervasive and exists in increasingly subtle and complex hierarchies. This concept was later elaborated by medieval and Renaissance thinkers into the Great Chain of Being, a schema to rank all life-forms from worms to angels on an ascending scale.

The Middle Ages was a volatile time when the Greek scientific spirit of Aristotle, Euclid, Democritus, Pythagoras, and Hippocrates contended with the old mythologies. The medieval Hermetics, or alchemists, exemplified the struggle. They mingled Gnosticism, Christianity, and theologies from Egypt, Babylonia, and Persia. They believed in creation from a preexisting chaos that included the grotesque and irrational. They thought of mutability, darkness, and mud as life-producing, of descents into chaos and encounters with monsters as revitalizing, and of creation as an ever-renewing process. Their dictum, in common with the astrologers, was "as above, so below." Yet the alchemists were also scientists who worked with scientific instruments and methods and made important chemical discoveries.

By the time of Galileo, Kepler, Descartes, and Newton, the scientific spirit and its suppression of chaos had gained the upper hand. Newton's laws of celestial mechanics and Descartes' coordinates (which allowed scientists to envision the universe as a huge grid) made it seem that everything could be described in mathematical or mechanical terms.

In Napoleon's time, the French physicist Pierre Laplace could reasonably imagine that scientists would one day derive a mathematical equation so powerful it would explain everything. The Yellow Emperor, carrying the wand of reductionist science, had cast his spell. Disorder was imprisoned and forced to reflect the gestures of a universal order. Just how did this occur?

Essentially reductionism is a watchmaker's view of nature. A watch can be disassembled into its component cogs, levers, springs, and gears. It can also be assembled from these parts. Reductionism imagines nature as equally capable of being assembled and disassembled. Reductionists think of the most complex systems as made out of the atomic and subatomic equivalents of springs, cogs, and levers which have been combined by nature in countless ingenious ways.

Reductionism implied the rather simple view of chaos evident in Laplace's dream of a universal formula. Chaos was merely complexity so great that in practice scientists couldn't track it, but they were sure that in principle they might one day be able to do so. When that day came there would be no chaos, so to speak, only Newton's laws. It was a spellbinding idea.

The nineteenth century tested the spell severely, however. For example, even as early as the mid-eighteenth century, scientists had begun to scratch their heads over why they couldn't invent a perpetual motion machine. Maddeningly, they found that whenever they ran a machine, some of the energy fed into it turned into a form that couldn't be recovered and used again. The energy had become disorganized, chaotic. This progressive disorganization of useful energy led to the important idea of entropy and the founding of the science of heat, thermodynamics.

For a while entropy challenged the notion of universal Newtonian order. Did the fact that a machine constantly needed new energy and that all forms are inevitably doomed to be crushed under the heel of accumulating entropy and decay mean that chaos is a principle as powerful as that of order?

In the 1870s Viennese physicist Ludwig Boltzmann attempted to neutralize the challenge of entropic chaos by proving that Newtonian mechanics was still universally true on the reductionist level of atoms and molecules. The movement of those cosmic watch parts always obeyed Newton's laws, Boltzmann argued, but in a complicated system where trillions of atoms and molecules are careening around colliding with each other, it becomes less and less likely that they'll all stay in an ordered relationship. In the grand scheme of things, ordered arrangements of large groups of atoms and molecules are highly improbable. So it's not surprising that when such ordered relationships do occur, they will relatively quickly break down. Boltzmann postulated that eventually even the atomic structure of our solar system would be shuffled into mere randomness. Reductionists now imagined that the end of the universe would be a state of general homogeneity, a lukewarm molecular cosmos: meaningless, sexless, formless.

However, to nineteenth-century scientists Boltzmann's precise definition of chaos was far different from the formless no-thingness, the active chaos envisioned by ancient myth. Mythic chaos had been "the first of all things" out of which bloomed forms and life. The passive chaos of entropy was the reverse. It was what happened when forms and systems ran down or ran out of the energy that had bound them together. The watch parts fell asunder and jumbled up, rebounding off of each other according to classical laws.

By introducing probability into physics, Boltzmann had saved reductionism from being corrupted by chaos, proving that the passive chaos of thermal entropy was simply an expression of the Newtonian order. The reductionist spell persisted.

About the same time Boltzmann was exposing the mechanics of entropy, Charles Darwin and Alfred Russel Wallace were announcing a theory explaining how new forms of life appear. Like Boltzmann, Darwin and Wallace saw chance—probability—as a key factor in the mechanistic processes governing complex forms. But here rather than mixing up complex order and destroying it, chance caused variations in individuals of existing species. Some of these variations survived and led to new species.

As the nineteenth century closed, belief in reductionism and mechanism prevailed, but the price paid for this was high. Humankind now saw itself as the product of an improbable collision of particles following indifferent universal laws. Dethroned as offspring of the gods, humans reenthroned themselves as the possessors of knowledge about those laws. By knowing the laws, it was thought, we would learn with increasing deftness to predict and control the entropy that afflicted complicated systems. In practical terms, passive entropic chaos couldn't be eliminated, perhaps, but it could be minimized or circumvented by an increasingly precise understanding of the universal mechanistic order underlying it.

The ancient Babylonians had envisioned many faces to chaos. Nineteenth-century reductionist science had disguised the chaotic face of entropy. It also masked another face of chaos by using a trick of reductionist mathematics.

Nineteenth-century engineers building their new bridges and steamships and other technological marvels repeatedly encountered disorder in the form of abrupt changes that were quite unlike the slow growth of entropy described by Boltzmann and the science of thermodynamics. Plates buckled and materials fractured. These phenomena challenged the powerful mathematics that had forged the Newtonian revolution.

For science, a phenomenon is orderly if its movements can be explained in the kind of cause-and-effect scheme represented by a differential equation. Newton first introduced the differential idea through his famous laws of motion, which related rates of change to various forces. Quickly scientists came to rely on linear differential equations. Phenomena as diverse as the flight of a cannonball, the growth of a plant, the burning of coal, and the performance of a machine can be described by such equations, in which small changes produce small effects and large effects are obtained by summing up many small changes.

A very different class of equations also exists, and nineteenth-century scientists were remotely acquainted with them. These were the nonlinear equations. Nonlinear equations apply specifically to discontinuous things such as explosions, sudden breaks in materials, and high winds. The problem was that handling nonlinear equations demanded mathematical techniques and forms of intuition then unavailable. Victorian scientists could only solve the simplest nonlinear equations in special cases, and the general behavior of nonlinearity remained wrapped in mystery. Fortunately, nineteenth-century engineers didn't need to penetrate that mystery in order to accomplish their mechanical feats, because for most of the critical situations they had to deal with, "linear approximations" could be used. Linear approximations are a version of differential equations. They rely on familiar intuitions and the tried and trusted reductionist links between cause and effect. Thus these equations were a kind of trick which masked the abrupt form of chaos. Once again scientists had kept the old reductionist spell in place.

The spell remained until the 1970s, when mathematical advances and the advent of the high-speed computer enabled scientists to probe the complex interior of nonlinear equations. As a consequence, within a few short years, this curious mathematics became one of the two winds driving the turbulent science.

THE NONLINEAR DEMONS

Nonlinear equations are like a mathematical version of the twilight zone. Solvers making their way through an apparently normal mathematical landscape can suddenly find themselves in an alternate reality. In a nonlinear

Figure P.2. A portrait of complex behavior over time is embedded in this computer plotting of a nonlinear solution. Is this a modern image of the primordial waters of chaos?

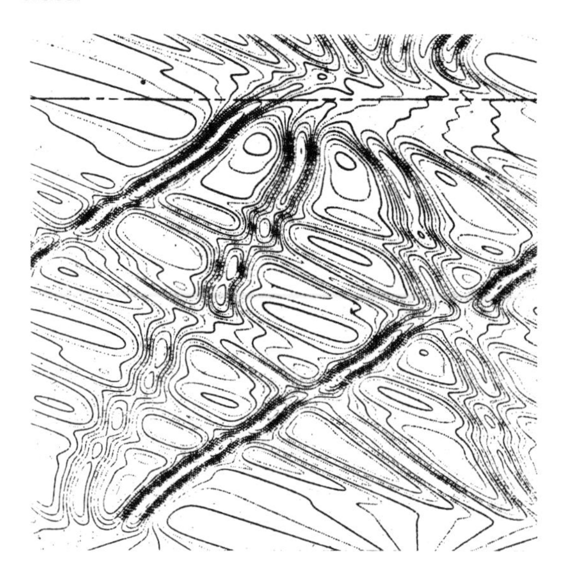

equation a small change in one variable can have a disproportional, even catastrophic impact on other variables. Where correlations between the elements of an evolving system remain relatively constant for a large range of values, at some critical point they split up and the equation describing the system rockets into a new behavior. Values that were quite close together soar apart. In linear equations, the solution of one equation allows the solver to generalize to other solutions; this isn't the case with nonlinear equations. Though they share certain universal qualities, nonlinear solutions tend to be stubbornly individual and peculiar. Unlike the smooth curves made by students plotting linear equations in high school math classes, plots of nonlinear equations show breaks, loops, recursions—all kinds of turbulence.

Nonlinear equations can be used to model the way an earthquake erupts when two of the vast plates that cover the earth's crust push against each other, building up irregular pressure along the fault line. The equation shows how for decades this jagged pressure mounts as the subsurface topography squeezes closer, until in the next millimeter a "critical" value is encountered. At this value pressure pops as one plate slips, riding up on the other and causing the ground in the area to shudder violently. Following the initial quake, instabilities in the form of aftershocks continue.

While nonlinear equations elegantly model such chaos and give scientists deep insight into the way such complex events unfold, they do not allow researchers to predict exactly where and when the next quake will come. As we'll learn, this is because in the nonlinear world—which includes most of our real world—exact prediction is both practically and theoretically impossible. Nonlinearity has dashed the reductionist dream.

The equations of Einstein's general theory of relativity are essentially nonlinear, and one of the amazing things predicted by the theory's nonlinearity is the black hole, a tear in the fabric of spacetime where the orderly laws of physics break down.

Cranking different values into nonlinear equations, systems theory scientists are able to picture the effects various policies and strategies would have on the evolution of cities, the growth of a corporation, or the operation of an economy. Using nonlinear models, it's possible to locate potential critical pressure points in such systems. At these pressure points, a small change can have a disproportionately large impact.

One difference between linear and nonlinear equations is feedback—that is, nonlinear equations have terms which are repeatedly multiplied by themselves. A growing awareness of feedback is the second wind driving the turbulent science.

GETTING LOOPED

In the late eighteenth century James Watt put a governor on his steam engine—and thus made a feedback loop. The most familiar governor feedback system is the one regulating the home furnace. The room cools down below a temperature set on the thermostat. The thermostat responds by switching on the furnace, which then heats up the room. When the room temperature slips above a second temperature set on the thermostat, the thermostat signals the furnace to shut down. The action of the thermostat affects the furnace, but equally the activity of the furnace affects the thermostat. Furnace and thermostat are bound in what is technically called a negative feedback loop.

Negative feedback loops show up in technology as far back as 250 B.C. when the Greek Ktesibios used one to regulate the height of water for a water clock. In the eighteenth and nineteenth centuries, governors were used widely. In the mathematical models developed in the 1930s to depict the relationship between predators and prey, negative feedback loops and other kinds of loops were implicit. The checks and balances of the U.S. Constitution have been found to act as negative feedback loops, and Adam Smith had them embedded in his descriptions of the "wealth of nations." But as systems scientist George Richardson of MIT says, "There is no evidence that the economists, politicians, philosophers, and engineers of the time pictured loops of any sort in their thinking."

It wasn't until the 1940s that negative feedback loops were recognized as such. Cybernetics, the machine-language information theory, made them popular. Then in the 1950s scientists began to take conscious note of the existence of feedback other than the negative kind. Positive feedback, for example.

Figure P.3

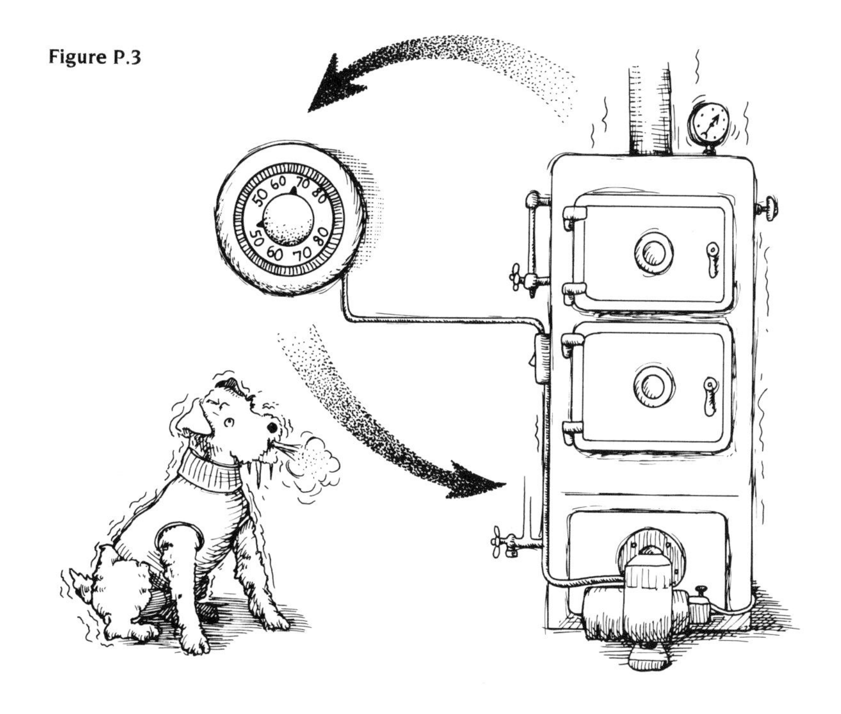

Figure P.4

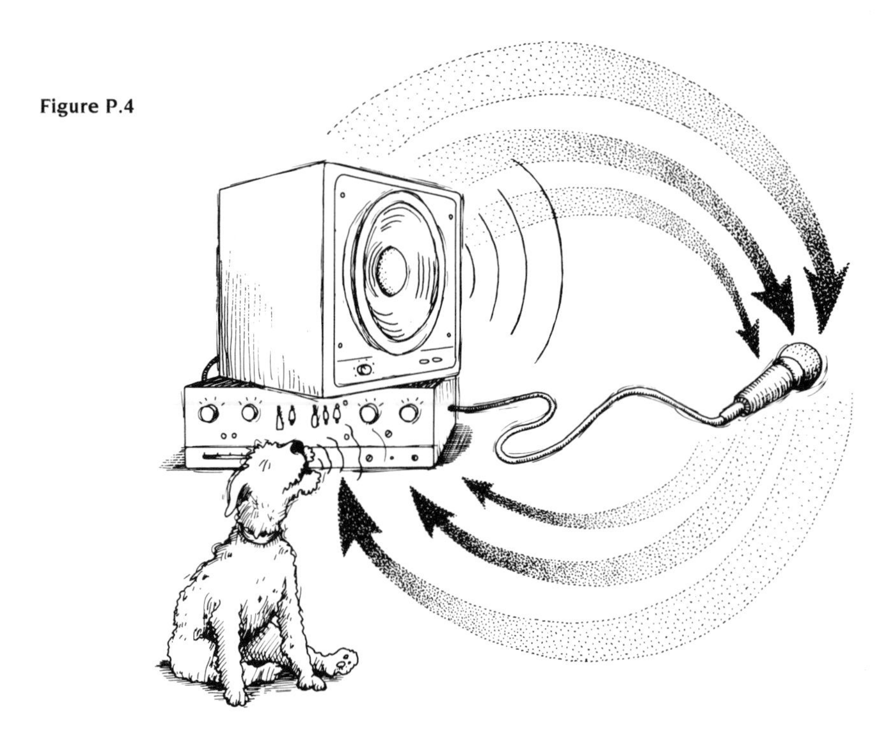

The ear-splitting screeches produced in a public address system are an example of positive feedback, which spurts into being when the microphone is placed too close to the loudspeaker. Output from the public address amplifier is picked up by the microphone and looped back into the amplifier, where it is then emitted by the speakers. The chaotic sound is the result of an amplifying process in which the output of one stage becomes the input of another.

Calling feedback "negative" and "positive" is not a value judgment. The names simply indicate that one type of feedback regulates and the other amplifies. It's now recognized that the two basic kinds of feedback are everywhere: at all levels of living systems, in the evolution of the ecology, in the moment by moment psychology of our social interaction, and in the mathematical terms of the nonlinear equations. Feedback, like nonlinearity, embodies an essential tension between order and chaos.

Through the recent exploration of feedback and nonlinearity, an ancient mirror-world has been rediscovered.

POINCARÉ'S PROBLEM: HOW NEWTON FELL AND NO ONE NOTICED

As it turns out, contemporary scientists weren't the first to rediscover this mirror-world. In the waning days of the nineteenth century, a brilliant French mathematician, physicist, and philosopher had already stumbled headlong into it and had sounded a warning. His cry was that reductionism might be an illusion. But though the cry was dramatic, it took almost a century before it was heard.

Henri Poincaré made his unsettling find in a field known as the "mechanics of closed systems," the epitome of Newtonian physics.

A closed system is one made up of just a few interacting bodies sealed off from outside contamination. According to classical physics, such systems are perfectly orderly and predictable. A simple pendulum in a vacuum, free of friction and air resistance, will conserve its energy. The pendulum will swing back and forth for all eternity. It will not be subject to the dissipation of entropy, which eats its way into systems by causing them to give up their energy to the surrounding environment.

Classical scientists were convinced that any randomness and chaos disturbing a system such as a pendulum in a vacuum or the revolving planets of our solar system could only come from outside chance contingencies. Barring those, pendulum and planets must continue forever, unvarying, in their courses.

It was this comfortable picture of nature that Poincaré blew apart when he impertinently wondered about the stability of the solar system. At first blush, the problem Poincaré posed seems rather absurd, just the sort of thing an ivory tower scientist might try to niggle over. After all, the planets have been around a long time, and at least since the era of the Babylonians it's been possible to accurately predict an eclipse years in advance. Wasn't the Newtonian revolution about this very point, the discovery of the eternal laws governing the movement of the moon around the earth and the earth around the sun? Moreover, Newton's laws were supreme to nineteenth-century physics. Knowing the law of force and the masses of the bodies involved in an interaction, all a scientist had to do to predict the effects of the interaction was solve Newton's equations. The law of force (the inverse square law of gravitation) was well understood and accurately measured.

All this was true, but Poincaré knew the palace secret: There was a small difficulty with the equations themselves.

For a system containing only two bodies, such as the sun and earth or earth and moon, Newton's equations can be solved exactly: The orbit of the moon around the earth can be precisely determined. For any idealized two-body system the orbits are stable. Thus if we neglect the dragging effects of the tides on the moon's motion, we can assume that the moon will continue to wend its way around the earth until the end of time. But we also have to neglect the effect of the sun and the other planets on this idealized two-body system. The problem is, and this was Poincaré's problem, that in taking the simple step from two to three bodies (for example, trying to include the effects of the sun on the earth-moon system), Newton's equations become unsolvable. For formal mathematical reasons, the three-body equation can't be worked out exactly; it requires a series of approximations to "close in" on an answer.

As an example, in order to calculate the gravitational effects of the sun plus the planet Jupiter on the motion of an asteroid in the asteroid belt (between Mars and Jupiter), physicists had to use a method they called "perturbation theory." The small additional effect that the motion of Jupiter would have on an asteroid must be added to the idealized two-body solution in a series of successive approximations. Each approximation is smaller than the one before, and by adding up a potentially infinite number of these corrections, theoretical physicists hoped to arrive at the right answer. In practice the calculations were done by hand and took a long time to complete. Theoreticians hoped they might be able to show that the approximations converge to the correct solution after adding just a few corrective terms.

Poincaré knew that the approximation method appeared to work well for the first few terms, but what about the infinity of smaller and smaller terms that followed? What effect would they have? Would they show that in tens of millions of years the orbits would shift and the solar system would begin to break apart under its own internal forces?

A modern version of Poincaré's question involves elementary particles being sped around the ring of a particle accelerator. Will the orbits of these particles remain stable or will they change unpredictably?

Mathematically, the many-body problem Poincaré was tackling is nonlinear. To the ideal two-body system, he added a term that increased the nonlinear complexity (feedback) of the equation and corresponded to the small effect produced by the movement of a third body. He then tried to solve the new equation.

As expected, he discovered that most of the possible orbits for two bodies are only slightly altered by the motion of the third body: A small perturbation produces a small effect, but the orbits remain intact. So far the results were encouraging. But what happened next came as a considerable shock.

Poincaré discovered that with even the very smallest perturbation, some orbits behaved in an erratic, even chaotic way. His calculations showed that a minute gravitational pull from a third body might cause a planet to wobble and weave drunkenly in its orbit and even fly out of the solar system altogether.

Poincaré had thrown an anarchist's bomb into the Newtonian model of the solar system and threatened to blow it apart. If these curious chaotic orbits could really occur, then the whole solar system might be unstable. Small effects of the planets as they circled around exerting their various gravitational pulls on each other might, given sufficient time, conspire to produce the exact conditions for one of Poincaré's eccentric orbits. Was it possible that in time the whole solar system could become chaotically unsprung?

Until Poincaré, chaos had been assumed to be an entropic infection that comes from outside a system, the result of external contingencies and fluctuations. But it now appeared that a system sealed in a box and left untouched for billions of years could at any moment develop its own instabilities and chaos.

Poincaré revealed that chaos, or the potential for chaos, is the essence of a nonlinear system, and that even a completely determined system like the orbiting planets could have indeterminate results. In a sense he had seen how the smallest effects could be magnified through feedback. He had glimpsed how a simple system can explode into shocking complexity.

The immediate import of Poincaré's discovery was to challenge the grand Newtonian paradigm that had served science for almost two centuries. His result should have sent a wave of activity through physics. As it turned out, however, nothing much happened because history went in another direction.

A few years after Poincaré's work, Max Planck discovered that energy is not a continuous substance but comes in small packets, or quanta. Five years after that, Albert Einstein published his first paper on relativity. The Newtonian paradigm was being attacked on several fronts. The next generations of physicists were occupied with plumbing the differences between the classical Newtonian view of nature and the view from the perspectives of relativity and quantum theory.

Quantum mechanics in particular swept across physics. One of the most successful theories in the history of science, it had made accurate predictions about a host of atomic, molecular, optical, and solid state phenomena. Scientists marshaled it to develop the nuclear weapons, computer chips, and lasers that have transformed our world. But it also brought troubling paradoxes. Physicists learned, for example, that an elementary unit of light can behave schizophrenically like a wave or like a particle, depending on what the experimenter chooses to measure. The theory also proposed that if two quantum "particles" are separated by several meters with no mechanism for communication between them, they will nonetheless remained correlated in some mysterious fashion. As recent experiments show, a measurement performed on one such particle is correlated instantly with the result of a measurement on its distant partner.

As we described in *Looking Glass Universe*, these paradoxes and others eventually had the effect of driving a number of scientists like David Bohm to theorize that the universe must be fundamentally indivisible, a "flowing wholeness," as Bohm calls it, in which the observer cannot be essentially separated from the observed. In recent years, Bohm and a growing number of other scientists have used the "koans" of quantum mechanics to challenge the long-held view of reductionism. Bohm theorizes, for example, that "parts" such as "particles" or "waves" are forms of abstraction from the flowing wholeness. In the sense that parts seem autonomous, they are only "relatively autonomous." They are like a music lover's favorite passage in a Beethoven symphony. Take the passage out of the piece and it's possible to analyze the notes. But in the long run, the passage is meaningless without the symphony as a whole. Bohm's ideas give a scientific shape to the ancient belief that "the universe is one."

No one could have guessed that Poincaré's results would lead in the same direction. In the tumult over quantum theory and relativity his discovery fell into the background. Small wonder, since even Poincaré had abandoned the ideas, saying, "These things are so bizarre that I cannot bear to contemplate them."

It was not until the 1960s that his investigations were disinterred from old textbooks and merged with new work on nonlinearity and feedback, entropy, and the inherent disequilibrium of orderly systems. These became the volatile elements of the new science of chaos and change—and have led to some stunning new perceptions into the mirror-worlds of nature's wholeness.

In the beginning there was Apsu the Primeval,
and Tiāmat, who is Chaos.

MYTHS OF THE WORLD

Chapter 1

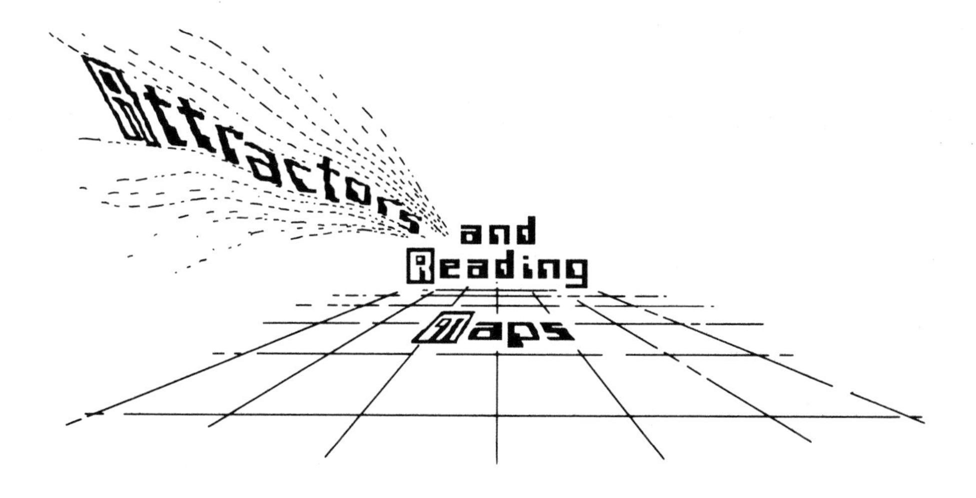

Then the Yellow Emperor breathed a sigh and said: "Deep is my error!"

LIEH-TZU

MAPS OF CHANGE

Our journey through the mirror-worlds of order and chaos begins on the side of the mirror where we will view from various vantage points what scientists have recently learned about the way chaos arises in orderly systems. The journey of these first four chapters will be a revisiting of the deep problem Henri Poincaré posed, but the perspective will be a fresh one. It will involve jabberwocky-like figures and Alice-in-Wonderland ideas.

The first of these odd figures is the attractor.

Attractors are creatures that live in a curious abstract place called "phase space." It's fairly easy to visit this space, but the trip requires a map. In fact, in the act of reading the "maps" of phase space and learning to identify the attractors that lurk there, we will pass from our familiar world of order to the very edge of the chaos Poincaré glimpsed. There, on that turbulent brink we will see nonlinearity and feedback throbbing in the form of an utterly wild and eerily beautiful beast called the strange attractor—but we're getting ahead of ourselves.

Let's start our trip by thinking about maps.

To find our way around in a new city, we use a street map; to drive through unfamiliar country, we use a road map. But there are many other kinds of maps: the highly stylized topological maps of the London Underground; weather maps showing winds, temperatures, and atmospheric pressures; maps that show the depths of rivers or the heights of mountains; maps in which the areas of countries are proportional to their populations or gross national product; maps of the electron density in a molecule, or of the spread of a new disease in Africa. Maps are imaginative pictures which allow thought to bring into focus aspects of reality that might otherwise be lost in details. With a good map we can appreciate some features of a reality we could otherwise miss, and we can explore this reality in a way that would be actually impossible without the map.

For example, hikers and climbers wanting to explore their reality and appreciate where they are use a map showing latitude, longitude, and altitude. Similarly, scientists wanting to explore the reality of a changing physical system—a dynamical system—make use of a "map" designed to visualize the dynamics, that is, the ways in which the system moves and transforms.

Suppose a scientist is interested in the changing movement (stoppings, slowings, and accelerations) of a car traveling from New York to Washington. Clearly it isn't enough to specify the location of the car at any one time; you need its speed. A scientist could make a graph showing these two aspects of the car's changing movement. Scientists call the imaginary "map" space in which the car's movement takes place, the system's phase space.

Phase space is composed of as many dimensions (or variables) as the scientist needs to describe a system's movement. With a mechanical system, scientists usually map the system's phase space in terms of position and velocity. In an ecological system, the phase space might be mapped as the population size of different species. Diagramming the movement of a system's variables in phase space reveals the curious byways of an otherwise hidden reality.

Let's set off a rocket ship and see what a phase space "map" looks like (Figure 1.1). Each point on the "map" is a snapshot of the rocket's height and speed (more accurately, its momentum, which is its mass multiplied by its speed) at one instant in time.

Between A and B the rocket surges up from the launch pad, its velocity increasing quickly. (In real life the acceleration may not be as

Figure 1.1

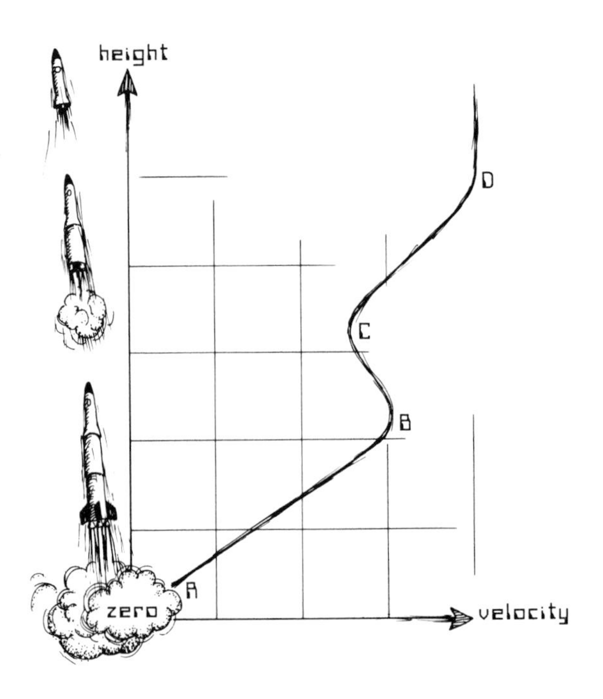

uniform as it is portrayed on the "map.") At B the first stage burns out and the rocket's acceleration sags a little from the effects of gravity. But at C the second stage cuts in and fires until D, when the rocket plucks itself free of the earth's downward pull and its velocity becomes constant.

As the illustration shows, a trip through phase space looks different from a trip through real space, just as a map of the London Underground looks different from the actual movement of subway cars through the tubes. Maps simplify reality in order to emphasize certain points. The rocket "map" is quite simplified.

To see how simplified, just consider the fact that our rocket ship is an object moving in three-dimensional space. For greater precision a scientist could try to capture that aspect of the rocket's movement in a more elaborate phase space diagram. Since a rocket can move in any one of three dimensions and can have —particularly when maneuvering in outer space—a different speed in any one of these dimensions, a rocket's phase space picture could be designed to have three space dimensions and three dimensions corresponding to each individual direction of velocity, making a $(3+3=)$ six-dimensional phase space.

The state of the rocket (its particular speed and position) at any moment is then given by a point in this six-dimensional phase space. The history of the rocket (how it has moved) is then given by a line, called a trajectory, in the phase space. Such multidimensional spaces are, of course, impossible to draw in our ordinary space. But scientists can draw a two- or three-dimensional cross section of a multidimensional space and mathematicians are quite happy to think about such higher spaces and determine their properties in abstract ways using fancy algebras.

In many cases, physicists investigate systems that contain several components, each one free to move in any of three directions with a different speed in each of three directions. Since a single particle requires a six-dimensional phase space (three space directions and three speed directions), a system of n particles will require a $6n$-dimensional phase space. For the moment we don't need to think about the exotic concept of $6n$-dimensional space. That's because while a rocket ship may theoretically require a very high dimensional space to describe it, in practice all the nuts and bolts, the gyros and other hardware, move at the same speed and maintain the same distances relative to each other. To describe the motion of the rocket ship we only have to think about the three directions in space and three directions of momentum.

This is commonly true of stable, orderly systems. Though they may ideally have a phase space containing a vast number of dimensions to move through, they actually settle down to move in a very tiny subspace of this larger space. The study of the movement of a system from order to chaos is, in a sense, the study of how this very simple and limited motion breaks down so that nature begins to explore all the implications of the much larger phase space at its disposal. The systems of nature are like animals that have been caged all their lives. Let out of the cage, they tend at first to move in a restricted way, not venturing too far, prowling around and around, performing a variety of repetitive movements. It's only when a slightly more adventurous animal breaks from this pattern that it gets out of sight of its home cage, discovers a whole universe to explore, and runs away in an entirely unpredictable way. As we'll soon see, nature's systems will often undergo rigid, repetitive movements and then, at some critical point, evolve a radical new behavior. It is these changes of behavior that the phase space maps help to clarify.

Figure 1.2

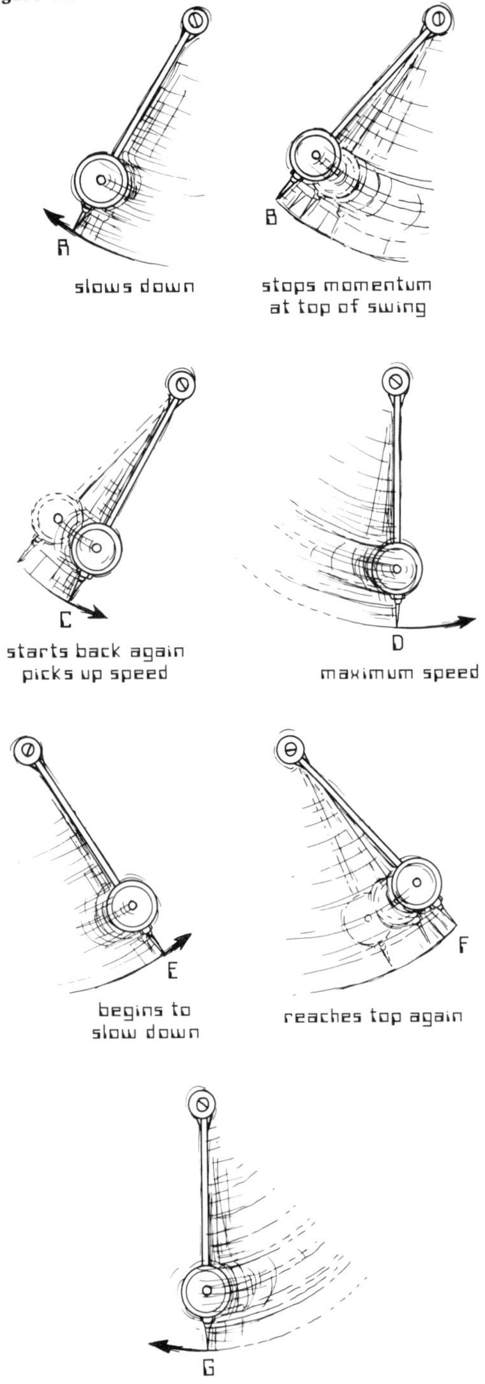

SYSTEMS THAT COME BACK TO THEIR CAGES

One of the simplest and most regular systems is one which acts periodically, that is, it returns to its initial condition again and again. A spring, violin string, pendulum, balance wheel on a watch, vibrating air column in an oboe, the output of an electric piano, day and night, pistons in an automobile engine, and voltage in a domestic AC electricity supply—all oscillate; they are all periodic.

These systems move back and forth, up and down, side to side so that with each complete oscillation they return to their initial position. It follows logically that the path of a periodic system must always return to the same point in phase space, no matter how complicated the returning path is. Such systems are well and truly caged.

A familiar example will illustrate these periodic systems: a pendulum ticking off the seconds (Figure 1.2). The pendulum swings up to the left, slowing down as it moves, until for an infinitesimal second at the top of its swing it comes to rest—and then returns, going faster and faster. It reaches its maximum speed at the bottom of its swing and, as it climbs to the right, begins to slow down again. The pendulum is one of the simplest systems exhibiting this periodic, repetitive behavior. In the absence of friction and air resistance, the pendulum should go on swinging forever.

Since the pendulum is confined to swinging back and forth in only one direction, scientists say, somewhat philosophically, that it has "one degree of freedom." The rocket, which is free to move in all directions of space, has three degrees of freedom.

Let's plot the path, or trajectory, of the pendulum on a phase space map. First, identify the top of the left swing at B. Here the momentum (mass multiplied by its speed) is zero and the pendulum is at the furthest part of its swing (maximum displacement). There's another point, F, at the right swing where the pendulum also has zero momentum.

Figure 1.3

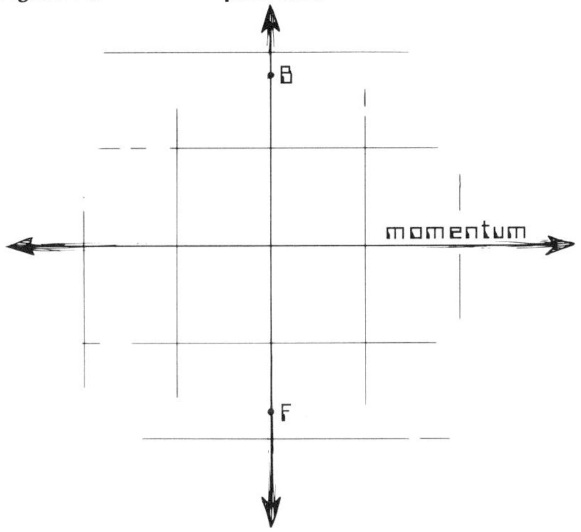

Now let's mark the two places where the pendulum is at its lowest point. Here its displacement is zero but its momentum (speed) is at the maximum. These points in phase space are D and G. At point D the pendulum is moving at maximum momentum to the right. At point G, it's moving with maximum momentum to the left.

Figure 1.4

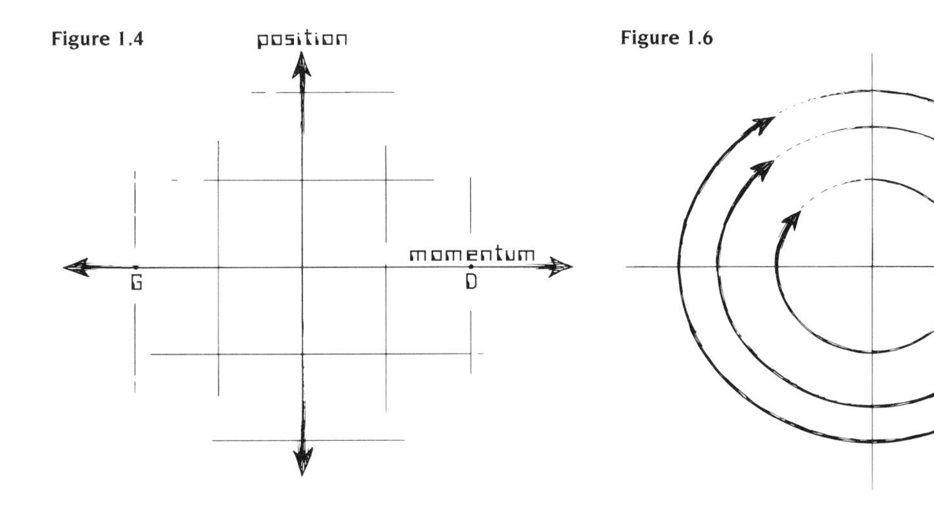

Finally, let's plot the phase space trajectory representing the entire motion of the pendulum for one cycle.

Figure 1.5

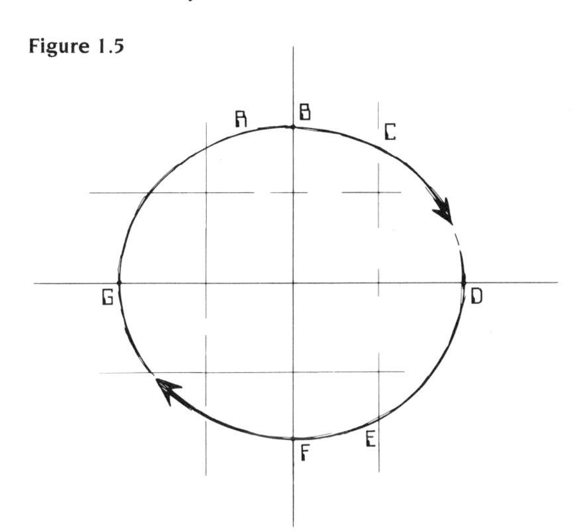

Since this plot repeats itself for cycle after cycle, the phase space map of a pendulum is a closed orbit.

If we give the pendulum a stronger push to begin with, then its maximum displacement will be bigger. In fact, on the same phase space map we can draw the same pendulum given different strengths of starting push.

Figure 1.6

Each of these circles represents a pendulum in a vacuum. But under ordinary circumstances pendulums become victims of friction and air resistance; they eventually slow down and stop unless there's a motor to keep them going. This process of a periodic orbit's decay can also be represented by a phase space map. The central point represents a pendulum with zero momentum and zero displacement —a pendulum at rest.

Figure 1.7

In fact, every earthly pendulum, no matter how great its initial displacement, will eventually come to rest at this final fixed point.

Figure 1.8

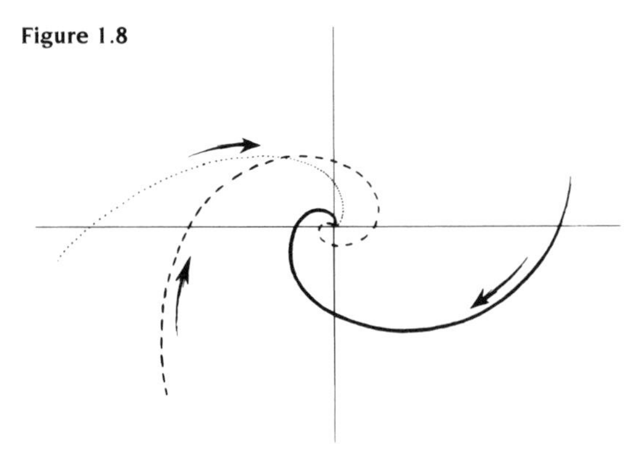

Because this point seems to attract trajectories to it, mathematicians call it either an "attractor" point, or a "fixed point attractor."

The attractor is a powerful concept that spans the mirror-worlds of both order and chaos. An attractor is a region of phase space which exerts a "magnetic" appeal for a system, seemingly pulling the system toward it.

Another way of getting a bead on this creature: Imagine a hilly landscape surrounding a valley. Smooth round rocks will roll down the hills to the bottom of the valley. It doesn't much matter where the rocks start or how fast they're rolling, all eventually end at the bottom of the valley. In place of the hills and valleys of a real landscape substitute hills and valleys of energy. Systems in nature are attracted to energy valleys and move away from energy hills.

Figure 1.9

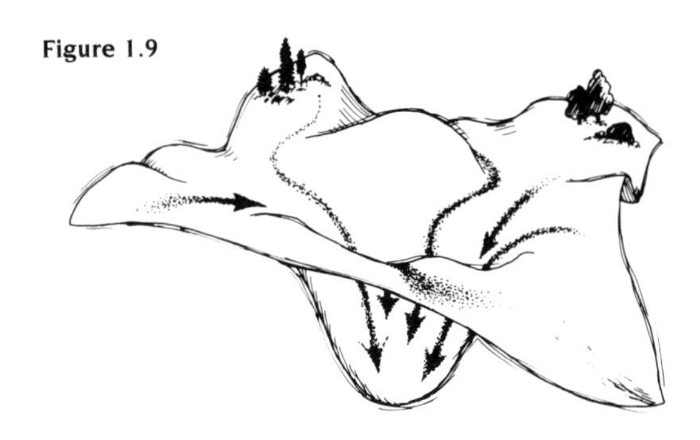

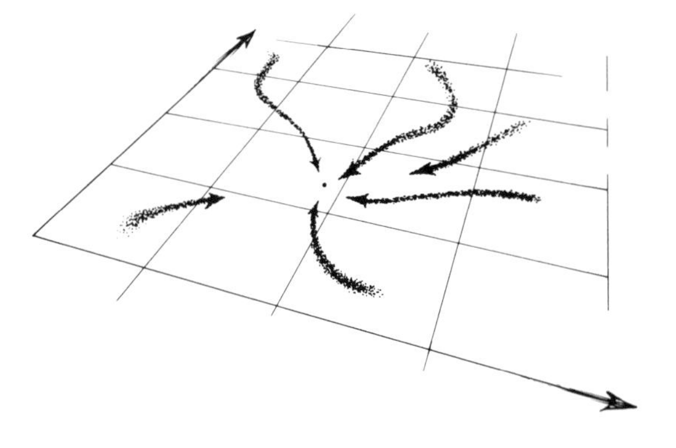

It's possible to have a landscape with two attractors—and a saddle between them. It's even possible to have a high-peaked mountain which acts as a point repellor. In such a landscape, phase space trajectories will avoid repellors and move toward attractors. In later chapters we'll see how scientists of chaos and change are visualizing fierce attractors full of folds, twists, and wrinkles more complex than the convolutions of the brain. But at the moment we're concerned with domesticated attractors which describe the evolution of systems in the classical world—systems where everything appears orderly. Step by step we will leave this world behind.

For example, let's return to the pendulum. In some modern clocks the pendulum is purely aesthetic because the clock is really driven by a more accurate quartz crystal. The electrical components inside the clock mechanism give the pendulum a periodic kick. So, the forces of friction and air resistance slow the pendulum down, but the periodic kicks speed it up. the result is that the pendulum swings at a regular rate despite the effects of friction and air resistance. In fact, even if a pendulum is given an additional push, or momentarily damped, it will eventually swing back to its original rhythm. This is clearly a new type of attractor. Rather than the pendulum being attracted to a fixed point, it is drawn toward a cyclic path in phase space. This path is called a limit cycle, or limit cycle attractor.

Figure 1.10

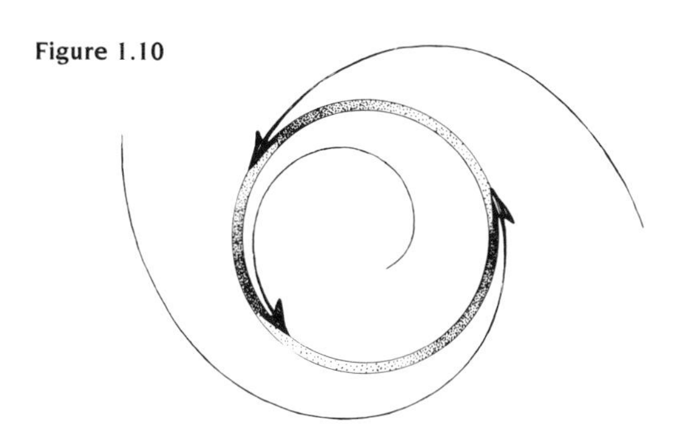

We should note that while a pendulum in a *vacuum* cycles without change, the pendulum's movement does not really involve a limit cycle, because the slightest perturbation causes the pendulum's orbit to change—to expand or contract a little. In contrast, a mechanically boosted limit cycle pendulum resists small perturbations. Try to let the system out of its cage and it comes running back home. The ability of limit cycles to resist change through feedback is one of the paradoxes discovered by the science of change. More and more, researchers are appreciating the way nature has of coupling continuously changing things together in order to end up with systems that effectively *resist* change.

An important instance of a limit cycle is the predator-prey system, an example of which showed up in the old records of the Hudson's Bay Company, a fur trading company in the Canadian north. Scientists noticed on the yellowing pages of the Hudson's Bay ledgers that over decades good and bad seasons for lynx and snowshoe hare pelts had followed a cyclic pattern which suggested that the population of these animals oscillated in a definite cycle. How could this be?

To understand it, let's follow the predator-prey system formed in a lake which has been stocked by trout and contains a few pike.

During the first year the delighted pike learn that they have an almost unlimited food supply of growing trout. The greedy pike flourish and breed so that as years go by the number of pike in the lake balloons—but at the expense of the trout.

At this point, with the pike's major source of food reduced, the lake becomes overpiked and these fish soon begin to die out.

Some years later, as the pike population drops, the trout multiply and again stock the lake. Consequently, the few pike now have plentiful food and their numbers once more rise. In this way, an oscillation between the number of pike and the number of trout, between predator and prey, sets up a cycle so that every few years the number of pike falls and the population of trout reaches a peak.

Scientists have closely studied this predator-prey system and shown that if you dump a load of trout into the lake anywhere in the cycle, the numbers will eventually settle back to follow the original limit cycle. Or if a disease kills off the trout, the population will spiral back up again to the cycle limits. A combined predator-prey system of pike and trout or lynx and hare is remarkably stable in its dynamics.

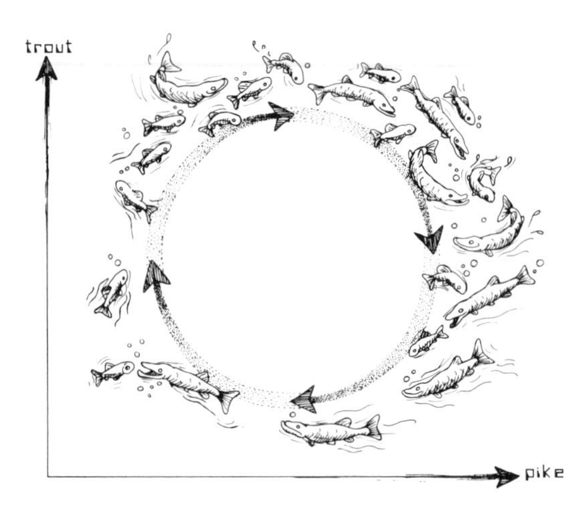

Figure 1.11

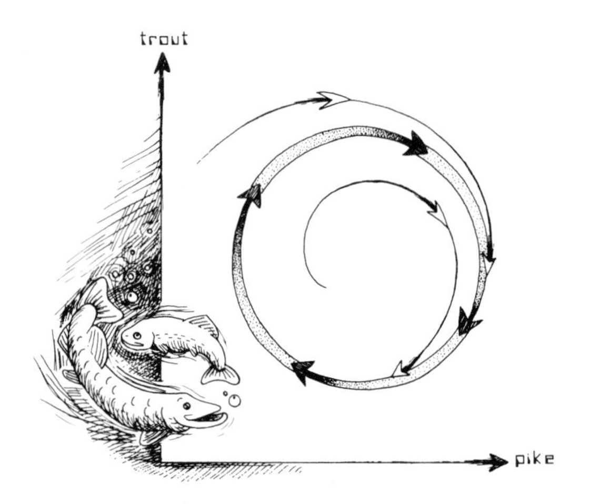

Figure 1.13. The spiral lines inside and outside the limit cycle indicate what would happen if you added trout to the lake or if a disease killed many off. After a time the system would return to the original cycle.

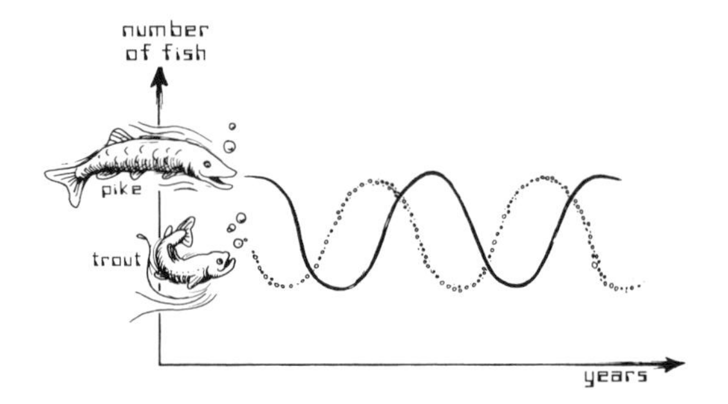

Figure 1.12

While the pendulum was a simple system, the predator-prey situation is considerably more complex. Here is a collection of many individuals, each one behaving randomly, yet all somehow creating a highly stable and organized system.*

* In fact, this kind of limit cycle stability is more than a little mysterious. How can individual random behavior produce such predictable structure? We won't get a full answer to that question until we cross to the other side of the mirror and see how order can emerge from chaos.

Limit cycles aren't always confined to a single periodicity. We can also have limit cycles describing the movement of the system with three variables, such as trout, pike, and anglers (Figure 1.14). This limit cycle is in a higher dimensional phase space.

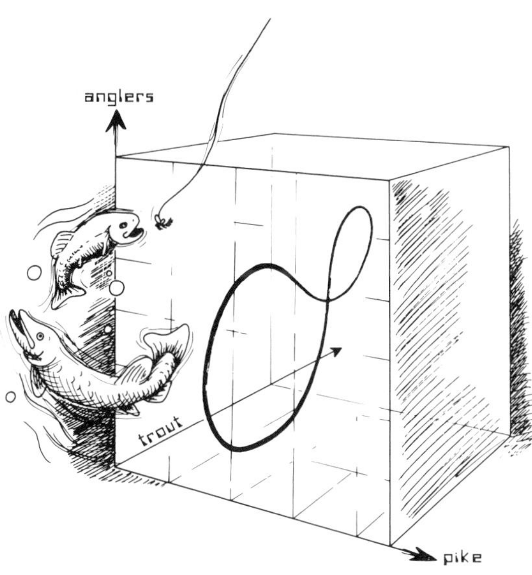

Figure 1.14

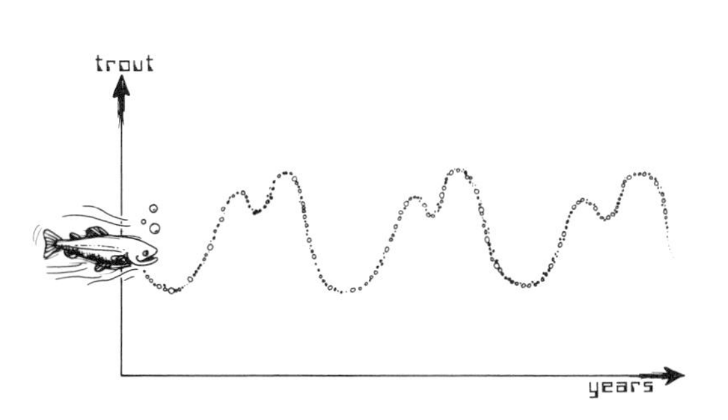

Figure 1.15. With a phase space built of three variables (trout, pike, and anglers), the limit cycle is more complex. Think of it this way: The number of trout is not only affected by the number of pike but also by the number of anglers who can prey on the fish. So the trout population in the lake can vary in two ways. Its limit cycle oscillates in two frequencies, as shown here.

We can also have two separate limit cycles interacting with each other. This often happens in electrical circuits and competing predator-prey populations. To visualize this kind of coupled limit cycle system, imagine the outputs of two different pendulums, A and B, each one with a motor. If we ignore pendulum A, then pendulum B's movement will have a simple limit cycle attractor.

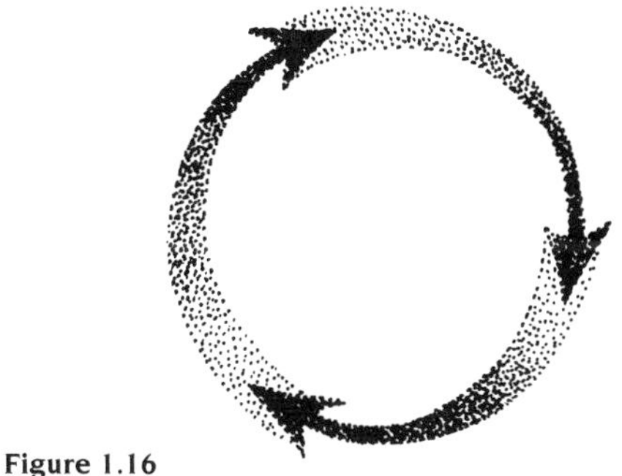

Figure 1.16

Likewise, if pendulum B is ignored, A's movement will have a simple limit cycle attractor.

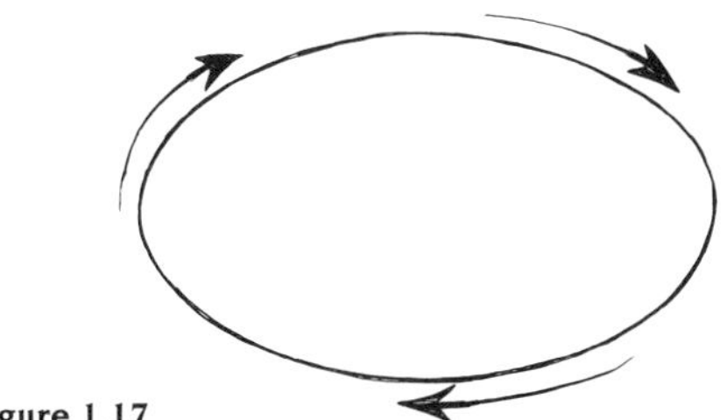

Figure 1.17

But if the two pendulums interact, the size of the phase space increases and the previously independent limit cycles become fastened together. It's as if cycle A is swept around in a circle by cycle B. The result of one circle being swept around by a second is the generation of a doughnut-shaped figure, which mathematicians call a torus. Here, instead of two interacting pendulums, we could also picture two interacting predator-prey systems. For example the trout-pike cycle might interact with an insect-frog cycle at the lake. Plotting the dynamics of this larger two-cycle system creates a torus attractor.

Figure 1.18

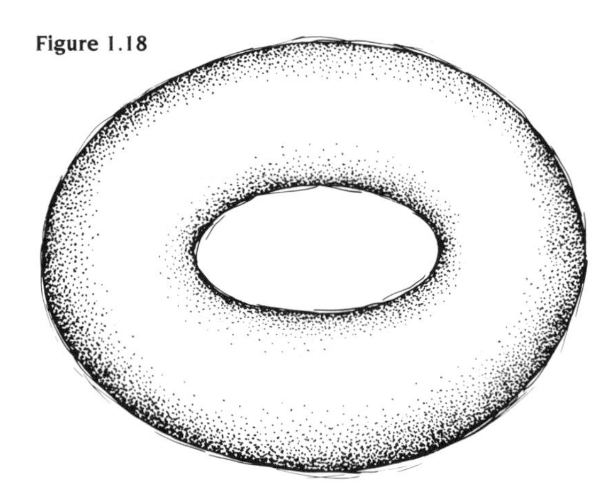

The torus attractor is a more evolved and complex creature than its limit cycle and fixed point attractor cousins. The state of a simple pendulum is described by a one-dimensional point that forms an attractor looping round and round in two-dimensional phase space. The combined state of two pendulums is described by a moving point that forms the two-dimensional attractor surface of a torus. The phase space inhabited by this twisted two-dimensional torus creature has three dimensions. But mathematicians are able to work with tori in any number of dimensions. That is, it's perfectly possible to couple together a whole toyshop of oscillators or a whole ecosystem of predator-prey relationships and to represent their combined motion on the surface of a multidimensional torus.

The torus is also handy for imagining a system with many degrees of freedom. What that means is: A simple pendulum or oscillator is free to move back and forth only in one dimension. But by loosening the pendulum's suspension system, it can swing from side to side as well, its full motion now in two directions. For physicists such an oscillating system, with two degrees of freedom, is the twin of two coupled one-dimensional oscillators: The oscillation of a two-degree-of-freedom system can also be described as a point moving on the surface of a torus. Tori in multidimensional phase space are just the thing for describing the orderly, apparently clockwork change that takes place in planetary systems.

The coupled motion of a pair of oscillators —whether they're planets or pendulums or predator-prey cycles—can be pictured as a line that winds around the torus, demonstrating that the surface of the torus itself is the attractor. Now let's dolly in on the torus for a closer look at this detail.

If the periods or frequencies of the two coupled systems are in a simple ratio—one twice as big as the other, for example—the twists around the torus join up exactly, showing that the combined system is exactly periodic.

Figure 1.19

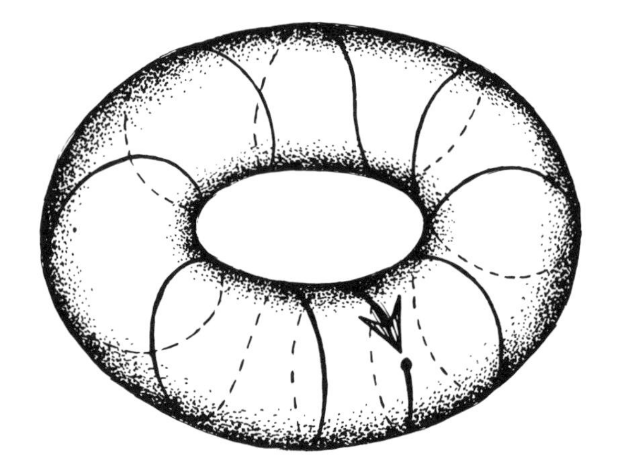

There's also another form of coupled oscillating behavior. Here the individual frequencies don't form a ratio, so they're what mathematicians call "irrational," which—as in the case of the words "positive and negative feedback"—is just a name, not a value judgment. Rational numbers like 1/2, 1/4, 3/4 and so on can always be expressed in terms of a finite number of decimals, 0.5, 0.25, 0.75, or as a simple recurring decimal, 1/3 = 0.333333. By contrast, an irrational number cannot be written down as a ratio and its decimal expression contains an infinite number of terms with no repeating pattern. The digits in an irrational number have a random order. In the case where the combined system forms an irrational frequency, the point in phase space representing the combined system will twist around the torus and never join up with itself (Figure 1.20). A system that looks almost periodic but never exactly repeats itself is called, quite logically, quasi-periodic. Mathematicians have proved that there is an infinity of rational numbers, but there is an infinitely larger infinity of irrational numbers, so on the face of things it would appear that quasi-periodic systems should dominate the universe.

Figure 1.20

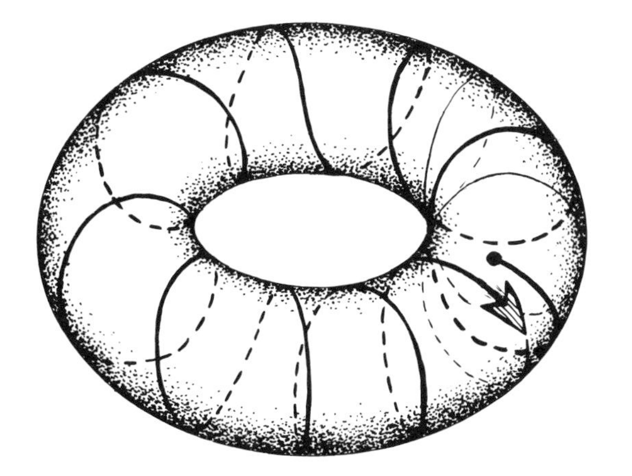

Scientists of the nineteenth century like Lord Raleigh and engineers of the twentieth century like Duffing and van der Pol studied a large variety of quasi-periodic systems that exhibit limit cycles around tori of various shapes. Such cycles were found by coupling together springs and pendulums, studying musical instruments, and calibrating the oscillations of electrical circuits.

At this point we notice that the kind of nature described so far by attractors is quite regular. Systems decay gently to fixed point attractors or oscillate in well-behaved limit cycle attractors around tori shapes. It is a classical world where scientists can predict the behavior of even quite complicated systems for long periods ahead. Scientists have also developed the notion of "asymptotic predictability"—meaning that even if they are ignorant about the exact position of a system at the moment, they are confident that no matter how far into the future they look, it will be moving somewhere on the surface of a torus and not wandering around randomly in phase space.

PROBING POINCARÉ'S POINT

But we've already seen how Poincaré threw a time bomb into prediction by finding a sort of black hole in Newtonian mechanics. Newton had shown how the motion of a planet around the sun, or of the moon around the earth, is an exactly solvable two-body, torus-shaped problem. But what happens, Poincaré asked, if you add to this description the effect of some additional planet? By stretching Newton's mechanics to three or more bodies, Poincaré found the potential for nonlinearity, for instability—for incipient chaos.

Poincaré's discovery was fully understood only in 1954, as a result of the work of the

Russian academician A. N. Kolmogorov, with later additions by two other Russians, Vladimir Arnold and J. Moser (the three being known collectively as KAM).

Before looking into what KAM discovered we should say that the kind of physics Poincaré had questioned is still being taught. Physicists still find it helpful to break up a complicated system in an abstract, mathematical way. So they mathematically reassemble the orbits of several planets or a bridge in high winds or a running engine into a set of simple oscillations, coupled together like a series of pendulums, and portrayed on a torus of some particular dimension.

Initially scientists believed that in theory they could do this sort of reductionist analysis for all complex systems. They were convinced that the corrections required to account for additional coupling oscillations would be small, and wouldn't affect the torus picture in any significant way. Poincaré's "bizarre" effects were exceptions where even the smallest additional term, the slightest gravitational pull of a third body, could spell the huge difference between a system exhibiting an orderly movement—confined to its torus—and becoming violently chaotic.

Did Poincaré's discovery imply that the whole universe is potentially chaotic, a fraction of a decimal point away from annihilation? KAM's answer was a resounding yes and no.

From their calculations they concluded that the solar system won't break up under its own motion provided that each of two conditions applies:

First, that the perturbation or influence of the third planet is no bigger than the size of the gravitational attraction of a fly as far away as Australia. Physicists hope that they can refine the KAM theorem to prove that larger-than-fly-size perturbations also won't affect the orbit (but they're still working on it).

The second condition preventing solar system disintegration is a requirement that the "years" of the planets in question don't lie in a simple ratio like 1:2 or 1:3 or 2:3 and so on. In other words, to remain stable, the planets must be quasi-periodic, the motion of their combined orbits looping around the torus again and again without ever joining up. In such cases the orbits will remain stable even under the perturbations of a third planet quite a bit larger than a fly.

But what happens when the planetary years coincide to form a simple ratio? Here the twisting system's path around the torus joins up, meaning that with each orbit the effect of the perturbation is amplified. The result is a resonance—analogous to the positive feedback in an amplifier in which small effects build up over time to produce a very large result, a screech of chaos. Mathematically this amplification causes the surface of the torus to blow apart in its phase space. The planet is still attracted to the surface and tries to reach it, and in the effort flops around chaotically until finally its orbit snaps and the planet flings off into space.

All this is according to the theory of Poincaré-KAM mathematics. Is there any evidence that such a mirror-world invasion of chaos into order actually takes place in our solar system's stately celestial mechanics?

Uncannily enough, when scientists looked they found gaps in the asteroid belt at exactly those places where the "years" of Jupiter and an asteroid would form a simple ratio. The gap indicates that any planet that happened to inhabit that orbit would rapidly shoot off into space.

Jack Wisdom of the Massachusetts Institute of Technology has scrutinized the latest results from the *Voyager* flyby and learned that many of the moons in the solar system must, at some time or another, have undergone a phase of chaotic motion, but then stabilized themselves by locating a quasi-periodic orbit. Hyperion, a glob-shaped tumbling moon of Saturn, appears to be in such a chaotic phase at the moment.

Wisdom has also applied the KAM theory to account for the meteorites that strike the earth. Scientists agree that these lumps of matter must originate in the asteroid belt. But how do they get to earth? By taking into account the combined gravitational influence of Jupiter and Saturn, Wisdom has shown how asteroids that stray into the resonance condition become subject to eccentric behavior that will eventually slingshot them toward us.

Gaps in orbits have also been noted in the rings of Saturn. Here the nonlinear (positive feedback) interaction is caused by Saturn's inner satellites. The gaps in the ring system correspond to simple ratios between the periods of rotation of the rings and the perturbing moons. This is evidence of both the relatively long-term stability of the rings and of the instability of certain of their orbits.*

And inside the instabilities are yet more surprises. When the gaps in planetary orbits like the asteroid belt or Saturn's rings are examined in detail the mathematics detects a pecularity of the mirror-world. There are gaps within gaps, like the cascade of reflections from an object placed between two mirrors.

In Saturn's rings, for example, the large-scale gaps between moons and rings are reflected on a smaller scale in the gaps between chunks of the ring material.

Mathematically this means that the torus breaks up into smaller and smaller tori. While some of these tori become stable, others don't. In the region between each torus lie smaller scale unstable orbits. Thus in regions where orbits form simple frequency ratios, the system reveals a gothic complexity.

In fact, the orbital situation we have just discussed provides our first glimpse into a new realization spreading across science—that randomness is interleaved with order, that simplicity enfolds complexity, complexity harbors simplicity, and that orders and chaos can be repeated at smaller and smaller scales—a phenomenon the scientists of chaos have dubbed "fractal."

Indeed, the solar system, physicists are beginning to see, is not the relatively simple mechanical clock pictured in Newton's day, but a system constantly changing, infinitely com-

Figure 1.21. Note the gaps where chaos pokes through the ring order.

* All this is enticing evidence for the KAM theory, but it should be emphasized that the question of Saturn's rings is quite complex and a number of theories are currently being tested by modeling on computers.

plex, and capable of unexpected behavior. So we return to Poincaré's problem. Does all this mean that even the solar system is capable of death throes and dying?

As it turns out, a little friction might be enough to cause that to happen.

It seems curious to think of the planets in terms of friction but the tides on earth dissipate the energy of the earth-moon system and a similar effect results from the friction between the dense, gaseous atmosphere of Jupiter and its moons. The frictional forces on the planets are thus very slowly changing the planetary and lunar orbits so that over millions of years they are gradually drifting. Possibly such movement is bringing them closer to regions of potential chaos. Is the solar system stable? Poincaré had asked. Unsettlingly, because of what modern chaoticians have discovered, his question must remain open.

However, if the solar system ever does break down and fall into chaos, and if there are any mathematicians around to observe it, they will at least know the cause. The culprit will be the Yellow Emperor's nightmare, a monstrous mirror-world creature utterly unlike the point attractor, limit cycle, or torus. Scientists have already recognized that this mirror-world attractor is inherently paradoxical. The systems which generate it jump around, they show no predictable pattern to their behavior. They are chaotic. Yet, as we'll eventually see, in their disorder is a shape. The attractor these systems cling to is a kind of organized disorganization of phase space—which is why scientists call it "strange."

Chapter 2

that Strange Attractor

The Yellow Emperor forgot his wisdom—all were content to be recast and remolded.

CHUANG TZU

LEONARDO'S DELUGE

In the nineteenth century chaos and regular order had little to do with each other, it was thought; they stood on opposite sides of the Yellow Emperor's mirror. But as Poincaré's insight has been enlarged by KAM and others, scientists are seeing that chaos is not merely a mindless jiggling, it's a subtle form of order. Our first example of this peculiar order was the chaotic asteroid eternally seeking its home in the structure of an attractor that has been fragmented across the phase space. Such a broken-up attractor has been dubbed "strange attractor"—a bizarre new object of mathematical analysis. (Figure 2.1)

It turns out there was nothing new about the strange attractor. Its presence had merely been hidden from us under another name—turbulence.

It's daunting to think of how many places turbulence occurs in nature: in air currents, in fast-flowing rivers swirling around rocks and the supports for bridges, in the way hot lava flows from a volcano, in weather disasters such as typhoons and tidal waves.

Turbulence often causes problems for humans. It interferes in our technology by jostling the movement of oil in pipelines; it jars the behavior of pumps and turbines, of trucks on highways, of ships' hulls in the water, and of the coffee in passenger jets' cups. The effects of turbulence in blood may damage vessels and lead to the accumulation of fatty acids on vessel walls; in the new artificial hearts, turbulence appears to have been the culprit causing the bloodclots that afflicted the first patients fitted with the device.

The turbulence that breaks up orderly systems and causes disorder to boil across our landscape in the forms of lava, wind, or water has long been an object of fascination for great minds. One of the earliest and greatest was Leonardo da Vinci, who made many careful studies of turbulent motion and became obsessed with the idea that a great deluge would one day engulf the earth.

Figure 2.1 A torus attractor fragmenting across space to create a strange attractor. Systems under the influence of a strange attractor bounce around chaotically following the attractor.

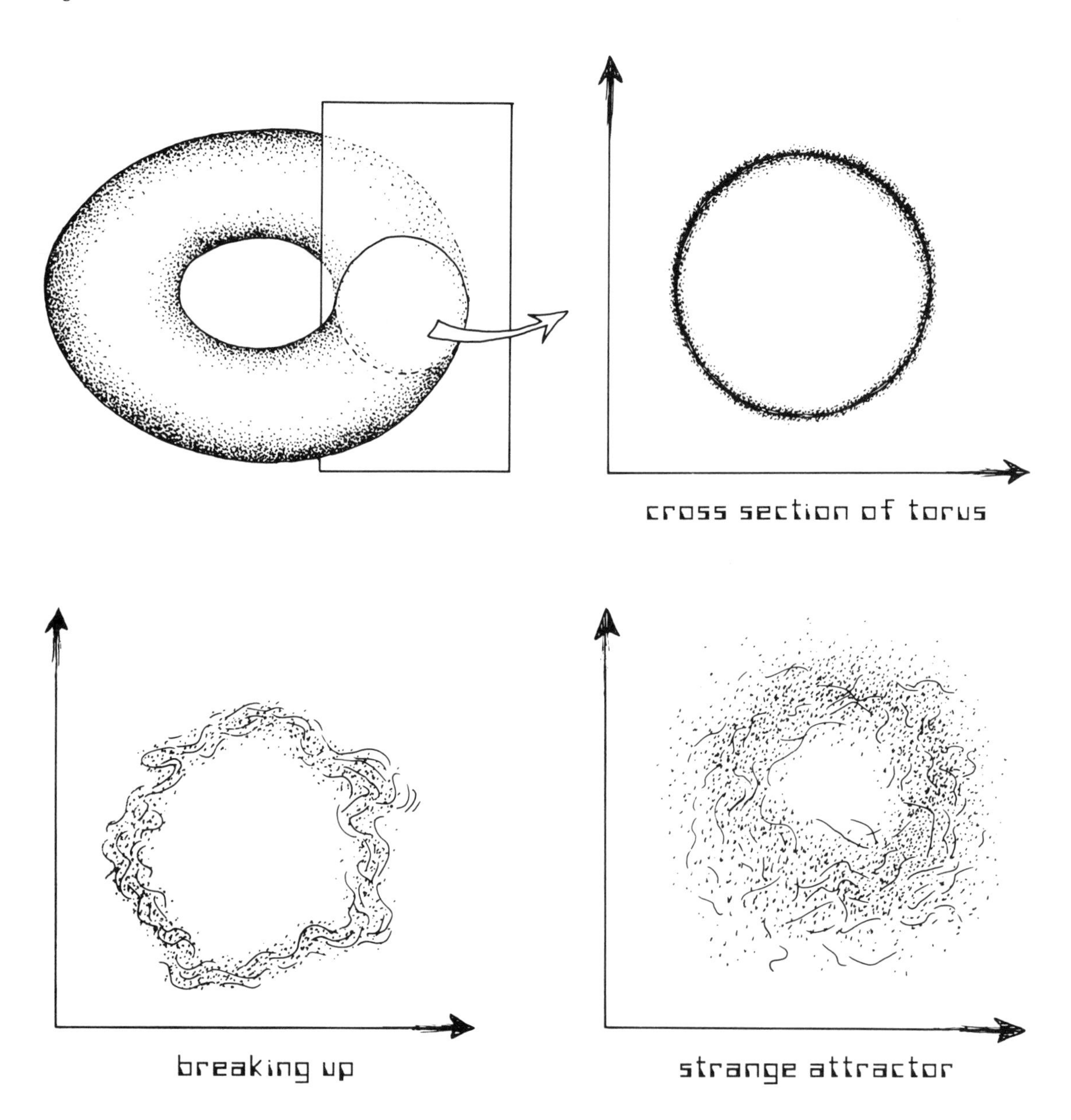

Leonardo avidly studied the flow of water in pipes and the eroding force of fast flow. In the nineteenth century turbulence drew the attentions of von Helmholtz, Lord Kelvin, Lord Raleigh, and a host of lesser known scientists who made important experimental contributions. But despite these efforts, turbulence essentially remained a backwater field of study. Dramatic results were hard to come by and the subject was largely opaque to science until recently, when it was recognized as a major area of research. A subset of the growing field of chaos theory, the study of turbulence focuses on the laws of attracting chaos in liquids and gases. Some scientists now think that turbulence (and chaos) may soon prove as important as quantum mechanics and relativity.

Part of the reason for recent interest in systems with so many degrees of freedom and such immensely complex dynamics is the spate of new sophisticated probes that make it possible to slice into a turbulent event and gather data about what is going on there. The development of superfast computers has allowed researchers to graphically display the byzantine results of nonlinear equations used to model turbulence. With visual displays, researchers can do slo-mos and instant replays of the processes tumbling about inside turbulent motion.

Still, the laws of turbulence have yielded only reluctantly to these efforts. Most of the progress made thus far involves descriptions of some of the routes that *lead* to turbulence.

Figure 2.2. One of Leonardo's studies of turbulent motion. The drawing depicts eddies within eddies within eddies. Larger swirls break up into smaller ones, and these again break up. Scientists call such a continual branching process "bifurcation."

A good place to start meditating on the problem of how turbulence arises is a river flowing slowly in the heat of summer.

The river encounters a large rock but divides easily and moves smoothly past the obstruction. If droplets of dye are put into the water, they produce flow lines streaming past

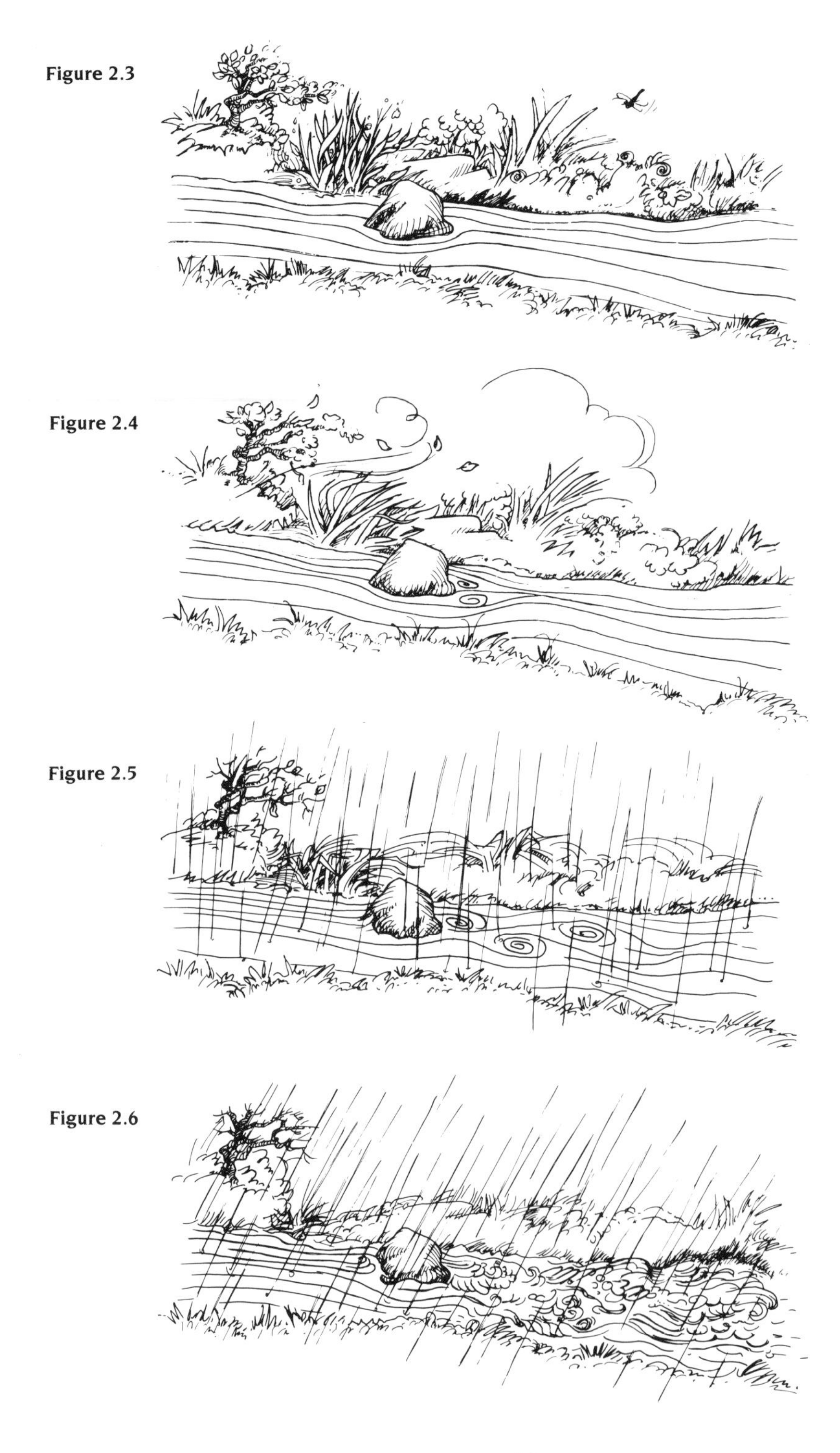

Figure 2.3

Figure 2.4

Figure 2.5

Figure 2.6

the rock, not diverging from each other or mixing up in any way (Figure 2.3).

With the coming of fall, rains begin and the river rolls along a little faster. Now vortices (limit cycles) form behind the rock. These are quite stable and tend to remain in the same place over long periods of time (Figure 2.4).

As the water's speed builds up, vortices detach and drift down the river, spreading the rock's disturbing influence further downstream. Earlier, a measurement of the river's flow rate downstream from the rock would have yielded a constant, smooth result. But now the flow rate fluctuates periodically as a result of the vortices (Figure 2.5).

As the river's speed picks up further, an observer sees the vortices unravel into local regions of choppy, swirling water. In addition to the periodic fluctuations of the water flow, there are now much faster, irregular changes: the first stages of turbulence (Figure 2.6).

Finally, with the water flowing rapidly, the region behind the rock seems to have lost all order, and measurement of flow rates in the region yields chaotic results. True turbulence has set in, and the motion of each tiny element of the water appears to be random. The region has so many degrees of freedom that it is beyond the powers of contemporary science to describe it.

In his observations and drawings of rapidly flowing water, Leonardo noted how vortices tend to fragment into smaller and smaller vortices, which then fragment again. The whole process en route to turbulence appears to involve endless divisions and subdivisions or bifurcations at smaller and smaller scales. Where do these bifurcations end? Is there a limit to their number? A fluid is ultimately composed of molecules; is it possible that true turbulence persists right down to the molecular level, or beyond?

The notion of vortices within vortices ad infinitum suggests that systems close to turbulence will look similar to themselves at smaller and smaller scales—suggesting again that the strange attractor of turbulence is a mirror-world.

A shiny sliver of this mirror was discovered in the nineteenth century by the British physicist Osborn Reynolds. By experimenting with pipes of different sizes, Reynolds was able to come up with a number—now called the Reynolds number—which tells an engineer just when the system will reach turbulence.

The Reynolds number is calculated by multiplying together several variables including the size of the pipe, the fluid's viscosity, and the rate of flow. Reynolds showed that as soon as the magic number is reached, turbulence appears. The critical number is one end of a spectrum that runs from smooth flow to vortices to periodic fluctuation to chaos. A curious feature of this spectrum is that it holds true at different scales. Using the Reynolds number scientists can simulate the complex movement of water in the Mississippi River on a tabletop. The flow of air around a model car subjected to a relatively slow airstream in a wind tunnel can mimic precisely the effects of a real car going at high speed on a highway. Amazingly, the approach of turbulence at a small scale reflects the onset of turbulence in the large. Reynolds had unknowingly encountered the curious self-similarity of the strange attractor.

TURBULENT DIMENSIONS

A Russian physicist was one of the first modern scientists to try to pin down the steps by which turbulence develops.

Lev Landau, who in 1962 became a Nobel laureate for his theory of superfluid helium, realized that turbulence begins progressively as the motions within a fluid become more and more complex. Much like Leonardo, he envisioned total turbulence appearing after a huge number of bifurcations had occurred.

Landau's theory received a boost in 1948 when German scientist Eberhard Hopf invented a mathematical model describing the bifurcations leading to turbulence.

In a smoothly flowing brook the various parameters that describe the flow are constant and unchanging. Even when the brook is disturbed by throwing in a rock, it soon settles back to its laminar flow. Since the variables defining the brook's flow don't change, the evenly flowing water can be represented by a single point in phase space, a point attractor. The point, in this case, represents the water's constant velocity.

Figure 2.7

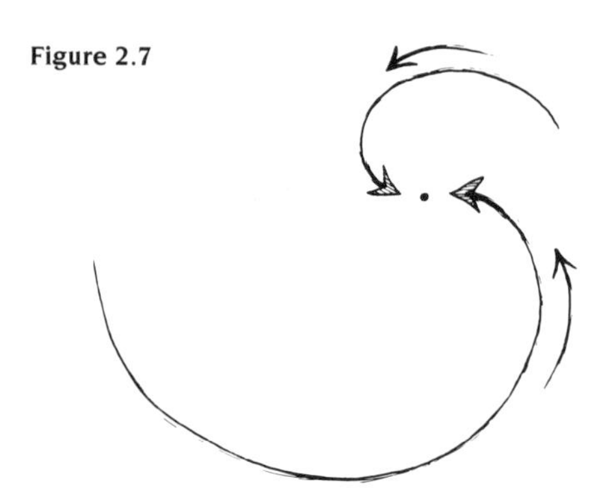

In a faster flowing brook, the smooth flow is warped by oscillations in which stable vortices form. Nevertheless, this flow is still highly regular and can be characterized as a single limit cycle. The perturbed brook will always return to the same basic oscillation, the same stable vortex, even if a rock is thrown in to disrupt it.

Figure 2.8

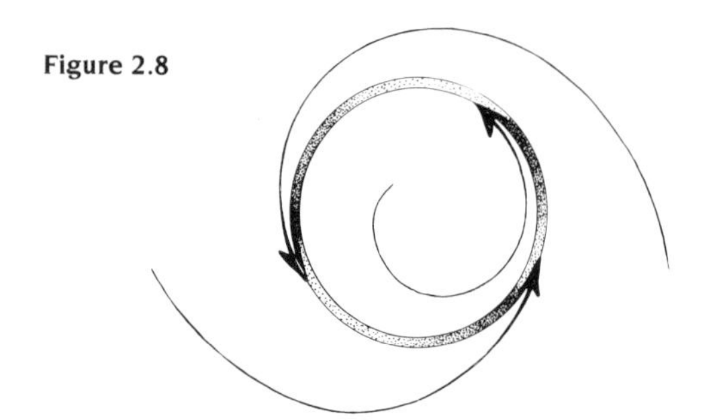

But such a description is almost paradoxical: When the speed of the brook is low, its motion is well described by a point attractor, but as the speed increases a limit cycle attractor applies. Clearly there must be some critical point at which the description of the brook's behavior jumps over from one attractor to the other. This critical point of instability is now called the Hopf instability.

Hopf went on to propose a cascade of further instabilities. The first instability involves a jump from point attractor to limit cycle. This is followed by a sudden charge to a torus attractor (a doughnut shape in three dimensions), then to a torus in four, five, six, and ever-increasing numbers of dimensions.

Hopf and Landau's picture is intuitively appealing; it recalls Leonardo's drawings of vortices within vortices. However, experiments have failed to confirm the higher dimensional toruses which are predicted by this model. Instead, observations of some systems indicate that though the beginnings of the transition from orderly to disorderly flow are the same as those described by Landau and Hopf, the system then takes a pathway to chaos that has even more amazing implications.

In 1982 a careful experiment was performed on the instability that appears in some convection currents when hot air rises from deserts or hot water curls and swirls up from the bottom of a pan. Researchers examining this particular instability, called the Bénard instability, found that turbulence set in much more rapidly than Hopf's hypothesis would suggest.

Physicist David Ruelle of the Institute des Hautes Études Scientifiques in France, with help from Floris Takens, created a new theory for this rapid onset of chaos.

Ruelle, who was the first to dub the attractor for turbulence and chaos with the name "strange," agrees with Landau and Hopf that in the convection current the smooth flow gives way to a first oscillation in which the point attractor jumps into a limit cycle. After this the limit cycle transforms into the surface of a torus. But Ruelle argues that at the third bifurcation something almost science fictional begins to happen. Instead of a system jumping from the two-dimensional surface of the torus onto the three-dimensional surface of a torus in four-dimensional space, it is the torus itself which begins to break apart! Its surface enters a space of *fractional* dimension. Put another way, the surface of the torus attractor is actually caught *between* the dimensions of a plane (two-dimensional) and a solid (three-dimensional).

To get some idea of what this means, consider a piece of paper, a two-dimensional object.* Crumple the paper up. The more tightly it's compressed, the more chaotic are its folds, and the closer the two-dimensional surface moves to becoming a three-dimensional solid. The Bénard convection is like the crumpling paper, or like a science fantasy character unable to choose between worlds. In a desperate fluctuating "effort" to escape to a higher dimension or return to a lower one, the current wanders in the infinite byways of "indecision" between the two dimensions and thus crumples up. The dimension this "indecision" inhabits is therefore not a whole dimension (not two-dimensional or three-dimensional) but a fractional dimension. And the shape the indecision traces is a strange attractor.

A striking experiment supporting Ruelle was devised by Harry Swinney of Haverford College and Jerry Gollub of the University of Texas at Austin (Figure 2.9). It involved studying the movement of a liquid between two cylinders. The outer cylinder is kept stationary while the inner one rotates. This sets up a flow in which different parts of the liquid travel at different speeds. With very low rotation speeds, the fluid flows uniformly. But as the rotation rate is increased, the first Hopf instability occurs. Now the fluid travels by means of a series of internal rotations like the twisting strands of a rope.

* Of course the paper is really three-dimensional, with one dimension being very thin. Nevertheless, at least metaphorically it is a fair approximation of a mathematical plane.

Figure 2.9

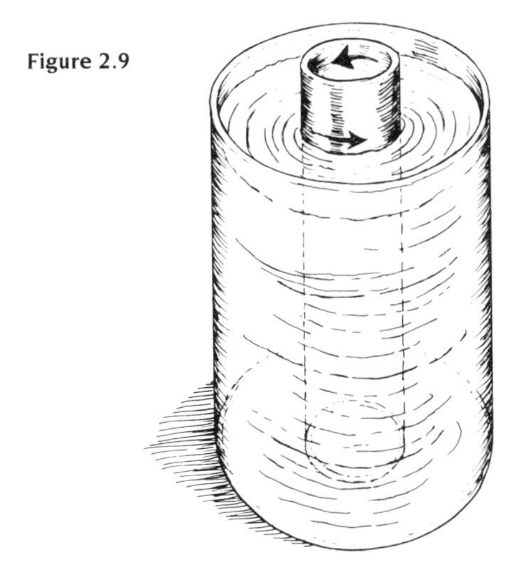

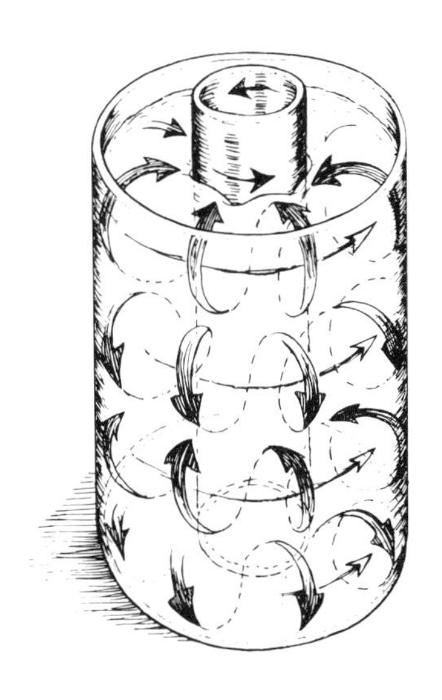

With the second Hopf bifurcation, a new set of internal rotations appears and the fluid twists with increasing complexity, oscillating at two different frequencies. When the rotation speed is further increased, the regular motion breaks into random fluctuations that, when plotted, wad themselves into the shape of a strange attractor with fractional dimension.

As scientists analyze the meaning of such experiments, they increasingly confront the irony of turbulence. Turbulence arises because all the pieces of a movement are connected to each other, any piece of the action depending on the other pieces, and the feedback between the pieces producing still more pieces.

Is the breakup of order into turbulence—that strange attractor—a sign of the system's infinitely deep interconnectedness? in fact, of its wholeness? Strange as it may seem, there is evidence that points in this direction.

Chapter 3

The Yellow Emperor said . . . "If we want to return again to the root, I'm afraid we'll have a hard time of it!"

CHUANG TZU

HOW THE WORMS TURN

Part of the evidence connecting wholeness, chaos, and the strange attractor comes from an occupation worthy of the characters in Alice's wonderland. By studying what happens when a simple mathematical equation is fed back into itself, scientists have wound deeply into the turbulent mirror. The study of such iterated equations has revealed a panoply of amazing mathematical properties, and it turns out that these properties—like Alice's mirror—reflect some of the seemingly crazy and convoluted changes that take place in our real world.

The growth of populations is a subject of interest to biologists, ecologists, epidemiologists—and to mathematicians as well. For behind the deceptively simple formulas of population growth lurks a rich and varied behavior that ranges from the simplest order to chaos.

History is replete with examples of populations out of control: the release of a small rabbit colony in Australia whose progeny exploded across the entire continent; the conquest of the northeastern United States by the gypsy moth caterpillar that escaped from a Boston laboratory; the migrating tide of killer bees; the waves of influenza which appear to lie dormant for years and then travel across the globe as pandemics, only to die down again before the onset of the next cycle.

Some populations multiply rapidly, others quickly die out; some rise and fall with a regular periodicity; others behave—as we're about to see—according to the laws of strange attractors, and chaos.

The growth of rabbit populations is too complex a starting point for understanding the onset of chaos. The reason for this is that some rabbits give birth while others are still coming to maturity or are carrying young. An equation that described rabbit population size would have to take all these factors into account.

A far simpler and equally illuminating population system to study is that of a parasite which lives during the summer and dies in the cold weather after laying its eggs. The gypsy moth is a good example. Let's begin with a small colony.

Assuming that about the same percentage of gypsy moth eggs hatch and survive each year, the size of a larvae colony this year is related to the number of larvae which metamorphosed into moths and laid eggs the year before. Suppose the size of a colony is 100 moths and the colony doubles each year. If the colony size is 200 for the second year, it will be 400 in the following year.

Figure 3.1

In the third year the colony size doubles again.

Figure 3.2

It's quite easy to give a general formula that allows the population in one year to be calculated from that in the previous year.

Figure 3.3

Of course not all populations double. Some may increase faster or slower. If we call the birthrate B, then each colony is B times bigger this year than it was last year. In our example of the gypsy moth we assumed $B = 2$, which led to population doubling. But now, with B taking other values, a variety of growths is possible.

Figure 3.4

This exponential growth equation works well enough for a very small or dilute population when there is plenty of food to eat and lots of empty space to expand into. But the formula is obviously limited. For example, applying it to rabbits breeding and doubling each generation projects that the original Australian pair should have spread out to cover the entire universe after only 120 generations! In the real world, exponential growth does not continue unchecked because any population system is dependent on other systems in the food chain. All these systems are interrelated, so the population size in the end depends upon the whole of the environment.

In 1845 P. F. Verhulst, a scientist interested in the mathematics of population growth, introduced a new term to describe the way a population develops in a closed area. In effect his term, which makes the growth equation nonlinear, was a simple and clever way to calculate the impact of all the other environmental factors on population expansion.

Before introducing this ingenious term, however, we need to do a little mathematical housekeeping. Up to now we've imposed no upper limit on the size of X_n (last year's population). But in order to be able to compare different populations and to make the calculation more regular, mathematicians perform a trick they call normalization. It's a useful way to compare different-sized populations. In essence the population is represented by a number that can vary between 0 and 1. $X_n = 1$ represents the maximum possible population, 100 percent. $X_n = 0.5$ represents half the value, 50 percent. It doesn't matter if we're talking about a population of several hundred moths or tens of thousands of bacteria. All we are interested in is comparing last year's population with this year's; that is, in looking at the ratios of population.

This trick of normalization, of allowing X_n, X_{n+1}, X_{n-1} to vary only between 0 and 1, has the effect of really simplifying the mathematics involved.

Now back to Verhulst's equation. In place of the simple growth equation

$$X_{n+1} = B X_n$$

he added an additional term, $(1 - X_n)$.

The right-hand side of Verhulst's equation now contains two competing terms, X_n and $(1 - X_n)$. As X_n grows, the $(1 - X_n)$ term diminishes. For a very small X_n, the $(1 - X_n)$ is very close to 1, so that Verhulst's equation looks just like the original growth equation. But what happens when X_n grows large, when it grows close to 1? Now the term $(1 - X_n)$ approaches

0 and causes the right-hand side of the equation to diminish—the birthrate falls (Figure 3.5). In other words, these two terms here work in opposition, one attempting to stretch out the population, the other squeezing it smaller.

Let's put it another way. Without Verhulst's term, the equation describes a process in which the population in any one year is proportional to that in the year before: The relationship is strictly linear. Multiplying X_n by the new term $(1 - X_n)$ can be written as

$$X_n - X_n \times X_n.$$

In other words, X_n is being multiplied by itself. Multiplying a term by itself produces feedback or "iteration" and nonlinearity. Growth from year to year now depends nonlinearly on what came before.

Verhulst's modified equation has a host of applications. It has been pressed into service by entomologists to compute the effect of pests in orchards and by geneticists to gauge the change in the frequency of certain genes in a population. It has been applied to the way a rumor spreads: At first a rumor will expand exponentially until nearly everyone has encountered it. Then the rate will drop off quickly as more and more people say, "I heard that one." Verhulst's equation also applies to theories of learning. What is learned now is related to the amount of information learned previously. Learning first increases, but after some time the learner becomes saturated so that more effort brings only minimal results.

The widespread application of the nonlinear version of the population equation has a surprising implication: In all the situations for which the equation applies skulks the potential for chaos.

Figure 3.5

NONLINEAR METAMORPHOSIS

To demonstrate the rich chaotic behavior of the iterated growth equation, let's begin with a population of gypsy moth larvae where some form of birth control has been imposed, for example spraying with insecticide. Assuming that the critters don't mutate, each year's population will shrink a little from that of the year before. If the birthrate B is 0.99, even a large starting population will eventually decline to zero. The colony expires.

But what happens when the birthrate is larger than 1, say 1.5? Because of the nonlinear Verhulst factor, a large population will at first decline but then settle down at a steady value of ⅔ or 66 percent of its original size. Likewise a very small initial population will grow to this same ⅔ limit.

With B (birthrate) equal to 2.5 the equation shows a slight oscillation as the two competing growth terms come into opposition, but, after that, the same steady population figure returns. It appears that the 66 percent figure has become an attractor.

Nudge B up to 2.98. Now what happens? The oscillation goes on longer but eventually the population settles into 66 percent of its original size—back to the attractor.

Push the value of B, the birthrate, further and these oscillations persist longer and longer, but the population eventually reaches a steady 0.66. However, when the birth rate reaches the critical value of 3.0, something new occurs. The attractor at 0.66 becomes unstable and splits in two. Now the population begins to oscillate around not one but two stable values (Figure 3.6).

Translated into real terms, this means that the small gypsy moth population breeds fanatically, leaving a large supply of eggs for the next season. But in the following season the whole region is overpopulated, creating a die-off, so the few surviving insects leave only a few eggs for the following year. The population

Figure 3.6

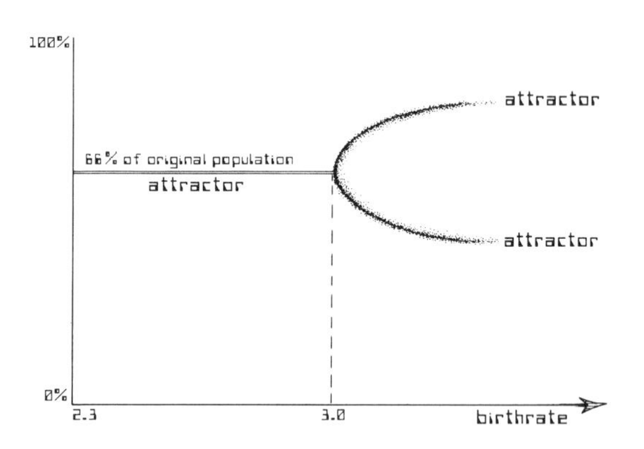

flips up and down between high and low values. The behavior of the system has become more complex (Figure 3.7).

As we crank up the birthrate above 3.4495, the two fixed points become unstable and bifurcate (branch) to produce a population oscillating around four different values. Now in each of four years the population of larvae is radically different.

When the birthrate reaches 3.56, the oscillations again become unstable, bifurcating into eight fixed points. At 3.596 another bifurcation, sixteen attractors this time. Things are rapidly becoming mazy. At this point it's almost impossible to see any order in the population of larvae rising and falling in your backyard. From year to year the number jumps in an almost random way and we can't discern the pattern. Finally, when the birthrate reaches 3.56999, the number of different attractors has increased to infinity.

Robert May, a Princeton physicist turned biologist, is one of the key figures in the story of how scientists learned about what is now called the "period-doubling route to chaos." A period is the amount of time it takes for a system to return to its original state. In the early 1970s May used a model based on the

Figure 3.7. A map of the first few bifurcating attractors for the nonlinear growth equation.

Verhulst formula that allowed him to increase or decrease the birthrate by altering the food supply. May found that the time it took for the system to oscillate back to its starting point doubled at certain critical values of the equation. Then after several period-doubling cycles, the insect population in his model varied randomly, just like real insect populations, showing no predictable period for return to its original state (Figure 3.8).

But, mathematically at least, that's not the end of the story. Scientists have learned that the period-doubling route to chaos contains a whole circus of previously unimaginable orders. Several are evident in Figure 3.9, a computer-generated plot of Verhulst's nonlinear birthrate equation.

The plot is a graphic display of the underlying structure of chaos, another image of the strange attractor.

First, notice the dark regions filled with points representing the virtual infinity of places the system may be found. In the birthrate range 3.56999 and 3.7 (between a and b on top of the plot) the system (yearly number of larvae) fluctuates unpredictably within four broad attracting regions and then two. These dark regions sweep toward each other until they meet where the b arrow is pointing. Here, at about 3.7, the population (the number of larvae in the backyard) could have almost any value, from very close to 0 to a very high figure (represented in the diagram by 1.0 in the upper left-hand corner of the graph), and from

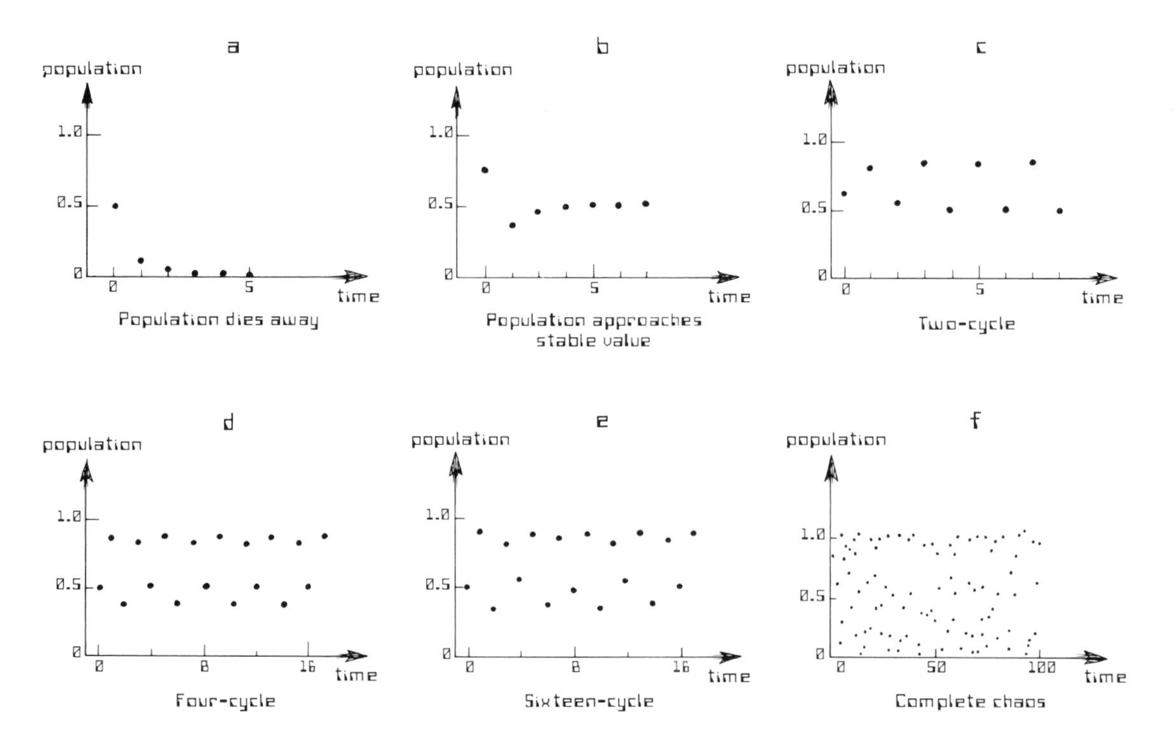

Figure 3.8. A breeding population shows the effect of a period-doubling route to chaos as food supply is varied. The population may very quickly die away (a) or reach equilibrium size (b). On the other hand, at certain critical values the whole system oscillates (c,d,e). Beyond another critical value for growth the population rises and falls chaotically (f).

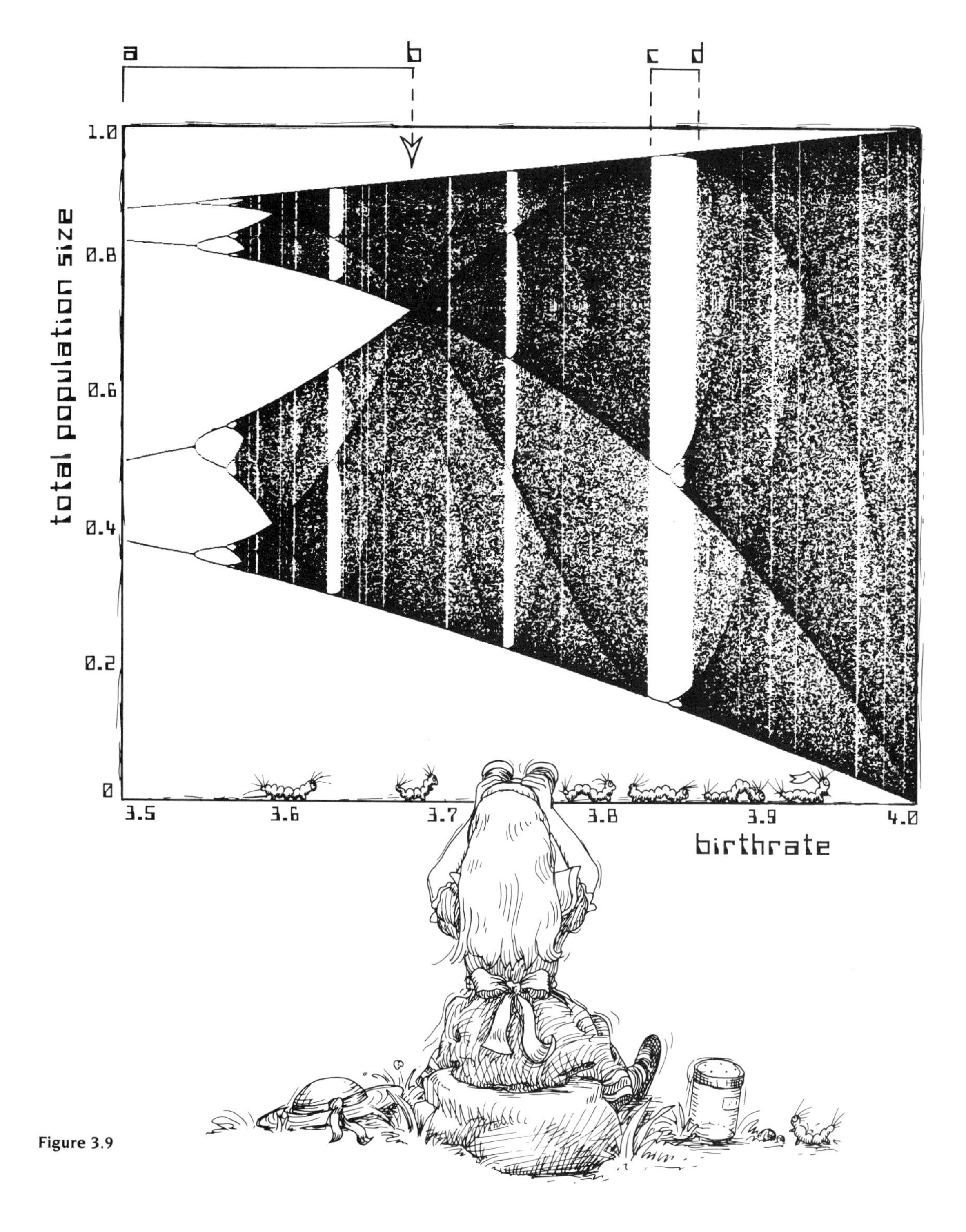
a
b
c
d
1.0
0.8
0.6
0.4
0.2
0
total population size
3.5
3.6
3.7
3.8
3.9
4.0
birthrate

Figure 3.9

year to year the population jumps in a crazy, unpredictable way. It's not, however, until the birthrate is 4.0 that the whole of phase space is filled. The way the whole plot fans out in the frame suggests that the chaotic filling of phase space is a strangely orderly process.

Second, notice how the dark lines form parabolas inside the spreading fan of chaos. These lines represent values where there is a higher probability of finding the system. Yet another form of order in chaos.

Third, notice the white vertical bands spaced throughout the expanding shadow of chaos. These are actually regions—or "windows" as physicists call them—where the system becomes stable. Around b = 3.8 (indicated by bracket c–d), for example, smack in the middle of all this spreading chaos, the population becomes predictable again and increases in two successive years and decreases in the third. But if the birthrate (food supply) is slightly bumped up, then the window jars open—chaos floods back in. These periods of stability and predictability in the midst of random fluctuation are called "intermittency."

INTERMITTENCY: THE CHAOS SANDWICH

You're relaxing, listening to the radio, when suddenly the music is interrupted by a salvo of static. It's not unusual for a short pulse of noise to interfere with reception on a radio or TV. This intermittent interference is often caused by an external source, for example, a neighbor's electric drill or an approaching thunderstorm. But it's also possible for intermittent noise to be generated within the circuitry of the amplifier itself. Japanese scientists have found that in superconducting switches—superefficient circuits where there is no resistance to the flow of electricity—intermittency will spatter up. If the current is increased through the switch, the average period between bursts of noise shortens. Conclusion: The switch is en route to chaos with no outside interference. The same phenomenon apparently struck a network of computers which a defense contractor, TRW, had strung in Europe. A report in the *New York Times* indicated that the network started to exhibit strange, unpredictable behavior. This also happened to a net of parallel processors put together by Xerox researchers who found their computers producing randomly different results from exactly the same calculation. The problem in these systems was not their design, engineers have concluded, but something inherent in the complexity of networks which contain nonlinear feedback loops. Some scientists believe that these observed bursts of intermittency reveal that massive computer networks such as those proposed for the Strategic Defense System ("Star Wars") or the high-tech monitoring of trading on Wall Street will always be subject to spasms of chaos. Chaos is like a creature slumbering deep inside the perfectly ordered system. When the system reaches a critical value the sleeping monster sticks out its jagged tongue.

Intermittency cuts both ways; it lives on both sides of the mirror. Think of it as islands of order in a sea of randomness or as the hiss of randomness interrupting the smooth broadcast of order. Intermittency could almost be thought of as a "memory" operating in nonlinear systems—the system's memory of its original limit cycle or periodic attractors. Iteration after iteration goes on as chaos (or order) moves through phase space. But in the intermittency regions the old order (or chaos) is discovered again momentarily and the very iterations producing the chaos (order) produce momentary regularity (or chaos).

Intermittency shows how the whole range of order from simple oscillations to the complexity of full chaos can be present in one system, with each extreme surfacing alternately. The phenomenon raises deep questions: To what extent do many different forms of order interweave in real systems? Are a system's simple orders and its chaos both features of one indivisible process? Intermittency is highly suggestive that this is the case.

An important form of intermittency is low-frequency noise. Not only an unwelcome defect in electronic amplifiers, this type of intermittency has been observed in the flow of current through metal and carbon films, semiconductors, vacuum tubes, diodes, and certain transistors. The voltage of electrical cells and convection currents in liquid are subject to short bursts of noise at low frequencies, and low-frequency intermittency is thought to be the source of disruption in nerve membranes. The length of the earth's day is also intermittent. Our day is the result of the planet's rotation around its axis, which should bring the sun directly overhead every twenty-four hours. However, there's a slight "wobble" in this regularity which takes place over a five-day cycle. Is this another example of chaotic noise intruding into the regular oscillations of the universe's nonlinear systems, a shadow of the interwoven complexity that lies behind apparently simple systems?

Think for a moment in reverse. Could intermittency be a reverse image of our place in the universe? We habitually see the cosmos from the point of view of order (that is, in terms of relatively simple orders). When our day wobbles or the radio spits static, we imagine those phenomena as disruptions of the structure that governs the universe we inhabit. But chaos theory suggests that a mirror-world point of view is also viable. We could imagine our familiar order as but an island of intermittency in the midst of a universe-large strange, or chaotic, attractor.

Figure 3.10. This is another kind of computer plot of a period-doubled system, showing islands of order amid a sea of chaos. The plot shows another facet of the strange attractor.

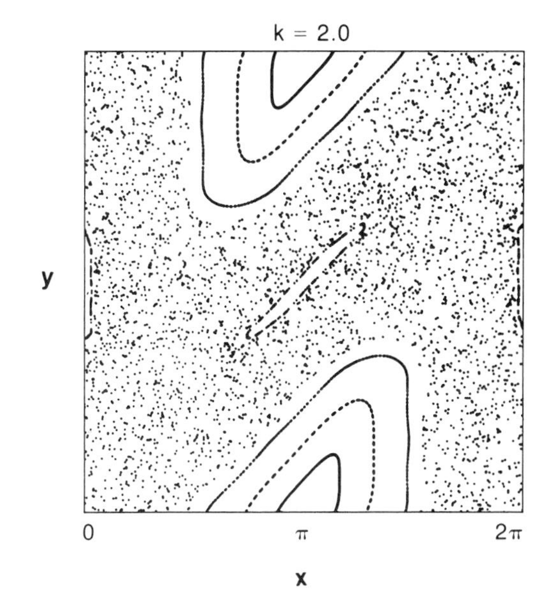

UNIVERSALITY

In the summer of 1975, while studying different period-doubling equations, physicist Mitchell Feigenbaum of the Los Alamos National Laboratory made a very significant discovery for chaos theory. Using a hand calculator, he tested a whole class of equations and found a universal scale to their period-doubling transformations. The equations he explored apply to such different phenomena as electrical circuitry, optical systems, solid-state devices, business cycles, populations, and learning.

Feigenbaum showed that the fine details of these different systems don't matter, that period doubling is a common factor in the way order breaks down into chaos. He was able to calculate a few universal numbers representing ratios in the scale of transition points during the doubling process. He found that when a system works on itself again and again, it will exhibit change at precisely these universal points along the scale.

Just as explorers were immortalized by having their names attached to the mountains or valleys they crossed, scientists leave their mark on the abstract landscape of nature's laws. The ratios Mitchell Feigenbaum uncovered have become known as the Feigenbaum numbers.

Armed with these numbers and the knowledge of period doubling, scientists all over the world soon began finding chaos everywhere.

At MIT, medical physicist Richard J. Cohen and his colleagues set up a computer simulation of heart rhythms and discovered that period doubling is a clue to the onset of a heart attack. In a normal heart, electrical pulses spread smoothly through the muscle fibers that force the heart's ventricle to contract and pump blood. When the muscle fibers are in their contracted state, they're impervious to electrical signals. Physicians call this period the refractory time. According to theory, variations in refractory time from one area of the heart's ventricle to another is the cause of fibrillation, the spasmodic twitching of a heart attack.

To test this theory, Cohen and his team varied the refractory times in their heart model and found that when a group of heart muscle fibers had a refractory time that was longer than the interval between heartbeats, trouble began. Because of their refractory time, these out-of-sync heart fibers could be stimulated to contract only on every other beat. As a result, electrical impulses from the contracting heart broke around these lagging fibers like water surging around a rock and causing turbulence. By slightly increasing the refractory times of a few fibers, the whole heart could be sent into period-doubling behavior until, past a critical value of refractory time, total heart muscle chaos set in.

At McGill University in Montreal, physiologist Leon Glass used a group of spontaneously beating cultured chick heart cells and stimulated them periodically. The result was that the time between regular beats doubled and then doubled again until reaching chaos.

Alvin Saperstein, a physicist at the Center for Peace and Conflict Studies at Wayne State University in Detroit, has made a preliminary study of the arms race leading up to World War II. He thinks the figures suggest the ratio of armaments between Nazi Germany and the Soviet Union went through a period doubling and was in the chaotic region when the war broke out. He emphasizes that his model is still very crude.

Period doubling has also been discovered in certain chemical reactions such as the Belousov-Zhabotinsky reaction, a combination of chemicals that appears to grow like a cellular life-form. Belousov-Zhabotinsky (as we'll see in *Chapter 3,* on the other side of the mirror) suggests that the route to chaos can simultaneously be a route to order.

It has now been shown that the growth of turbulence pictured by Leonardo can also occur by period doubling. In fact, Italian scientist Valter Franceschini confirmed Feigenbaum's numbers when he used a computer to analyze five equations that modeled turbulence in fluids. After discovering period doubling in 1976, Feigenbaum had been unable to get his papers on the phenomenon published because the editors of science journals found the concept too bizarre. Then in 1979 a colleague of Franceschini who knew about Feigenbaum's theory suggested the Italian scientist look for the Feigenbaum numbers in the equations he was studying. When Franceschini reran the calculations, the Feigenbaum universals leapt out.

Shortly afterward, two French scientists, Albert Libchaber and Jean Maurer, confirmed Feigenbaum's insight experimentally, though they were unaware of his work at the time. In their laboratory they discovered a symmetry to the chaos of the Bénard instability. They found it by very carefully heating liquid helium inside a 1-millimeter stainless steel box. Increasing the heating rate slowly and measuring the convection currents, the two researchers recorded a pattern of bifurcating oscillations that exactly followed the period-doubling route.*

The route of period doubling takes us deep inside the turbulent mirror; we catch a new glimpse of the strange attractor and are surrounded by a thicket of questions. How does period doubling really work? How does it produce (or reflect) the welling up of chaos and its expression of the apparent wholeness that exists between chaos and order? What is the strange attractor?

In part the answer to these questions lies in the phenomenon of iteration.

* The experiments were conducted with the liquid helium in rectangular containers. When the research was repeated by two German scientists using different-shaped containers, the route to chaos didn't involve period doubling. The apparent lesson is that there may be many as-yet-undiscovered routes to chaos.

Chapter 4

Uncle Lack-Limb and Uncle Lame-Gait were seeing the sights at Dark Lord Hill and the wastes of K'un-lun, the place where the Yellow Emperor rested. Suddenly a willow sprouted out of Uncle Lame-Gait's left elbow. He looked very startled and seemed to be annoyed.

"Do you resent it?" said Uncle Lack-Limb.

"No—what is there to resent?" said Uncle Lame-Gait. "To live is to borrow. And if we borrow to live, then life must be a pile of trash. Life and death are day and night. You and I came to watch the process of change, and now change has caught up with me. Why should I have anything to resent?"

CHUANG TZU

WHAT'S THAT AGAIN?

Iteration—feedback involving the continual reabsorption or enfolding of what has come before—crops up in almost everything: rolling weather systems, artificial intelligence, the cycling replacement of cells in our bodies.

Even in philosophy iteration has an important place. Consider the strange mental state induced by the philosophical iteration known as the "self-referent paradox." An early and famous example of this is the parable in which a man from Crete warns a passerby, "All Cretans are liars." Does this Cretan lie? If so, then his statement is false and all Cretans are not liars. But if he is telling the truth, then he, too, must be a liar. Notice how truth telling and lying swirl around each other, creating a kind of chaos and order in the brain.

A similar paradox may be unleashed into consciousness by a piece of paper containing on both sides the message: "The statement on the other side is false."

If a statement like this is given to a computer, the machine will stutter helplessly between "true" and "not true." In several "Star Trek" episodes Captain Kirk used self-referential paradoxes such as, "Prove that your prime directive is not your prime directive," in order to burn out the semiconductors of miscreant mainframes.

For a computer iterative paradoxes lead to chaos. For human beings they are said to have the opposite power—leading to creative insight or even enlightenment. In mystical systems like Zen Buddhism, self-looping koans supposedly set the mind of the student oscillating in a way that creates the conditions for it to bubble free into a new point of view (or a viewless point).

Figure 4.1

A famous Zen paradox cited by Douglas Hofstadter in his book *Gödel, Escher, Bach* involves two koans. The Zen master says one of them is true, though he doesn't know which the true one is. The koans are: (1) "A monk asked Baso: 'What is Buddha?' Baso said: 'This mind is Buddha.'" (2) "A monk asked Baso: 'What is Buddha?' Baso said: 'This mind is not Buddha.'"

As in the all-Cretans-are-liars paradox, a movement is set up where the mind's understanding of truth and falsity continually fold back on each other. The two koans (actually one koan) are mirrors of each other in the sense that one side is the reversed reflection of the other. Hofstadter says coyly that Zen masters have devised a way out of the mirror. Finding the exit entails the rather bizarre task of translating the two koans into pieces of starched string folded in on themselves according to definite rules (an appropriate image of the folding process of iteration). Some of these translation rules make the string more complex, some simplify it. Once all the folds are made, the Zen student sees which koan is true. However, in classic Zen fashion Hofstadter goes on to complicate matters by showing that it is also simultaneously impossible to find the true koan by using this folding method.

The logician G. Spenser-Brown has suggested that because a paradox constantly reenters itself, each iteration is like the tick of a clock. Such paradoxes, he believes, play the role of introducing time into logic, which includes the logic of mathematics and most of the important processes of thought. Some of those important processes involve language, which is itself a superlatively circular and self-referential device. Anyone who has tried to look up difficult words in a dictionary has a feel for this fact. For instance, the word *time* is defined in terms of words such as *period* and *instant*. But what do these words mean? Looking them up eventually leads back to the word *time*.

Self-reference also appears in biological systems, where the result may be Zen-like. At least that's what theoretical biologist Howard Pattee believes. Pattee thinks that while computers oscillate suicidally when trapped in a self-referential paradox, biological systems employ self-reference for their stability and may even use it to catapult themselves to higher forms.

Take bacteria, for example. These first forms of life on earth have no cell nucleus. They reproduce by dividing and making copies of themselves. Bacteria also have the ability to transfer among themselves—by a process that is not reproduction—bits of genetic material. This means that all the world's bacteria have access to each other's genetic storehouses. Through constant iteration of material in the genetic pool, bacteria are able to adapt to changing conditions very quickly. The downside of this biological form of self-reference, however, is that there are no real individuals among bacteria, just different strains of clones as the bacteria feed back into themselves, making copies.

Millions of years ago, nature may have used this form of self-referential paradox to great advantage as an efficient way of colonizing the planet with life. The disadvantage is that there is a limit to the intricacy of life-forms that can be made by this method. According to one theory (which we'll explore in *Chapter 3*), the successful iteration of bacteria overspread the earth and created chaotic conditions out of which sprang a new self-referential loop involving sexual reproduction. This elicited a new and incredibly dynamic level of evolutionary development.

In their book on microbial evolution, *Microcosmos*, Lynn Margulis and Dorion Sagan say that now a third loop is emerging. "In one of life's giant self-referential loops, changing DNA [which happened when sexual reproduction came into being] has led to the consciousness that enables us [through genetic engineering] to change DNA."

A number of theories in physics are proposing that at the smallest and presumably most basic level of matter, self-referential iterations also occur. Elementary particles generate themselves by a constant process of creation and destruction through iteration from the vacuum state. This means that the ultimate reductionist entity, the so-called building block of nature, owes its stability not to some rocklike permanence or static quantity, but to a dynamic cycling *quality* or process in which the particle constantly unfolds and enfolds within its quantum field.

Iteration suggests that stability and change are not opposites but mirror-images of each other. Consider the cells in your body. Every seven years or so they are completely recycled, iterated. The pancreas replaces most of its cells every twenty-four hours, the stomach lining every three days. Even in the brain 98 percent of the protein is recycled every month. Yet, though you are constantly changing, you remain essentially the same.

Like the wizard Merlin, who could appear in different disguises—a child, a bird, an old man—iteration performs its magic again and again in the science of change. Everything, from stability to chance to time, is generated by it.

MULTIPLYING THE DIFFERENCE

The honor of being the first to discern how iteration generates *chaos* goes to Edward Lorenz, an MIT meteorologist.

In 1960, Lorenz was using his computer to solve a number of nonlinear equations modeling the earth's atmosphere. Repeating one forecast in order to check some details, he plugged in his data on temperature and air pressure and wind direction, rounding off the figures in the equations to three decimal places instead of the six he had used in the previous run. He cranked the equation into the computer, and went out for a cup of coffee. When he came back he had a shock. The new result he saw on his screen wasn't an approximation of his previous forecast, it was a totally *different* forecast. The small, three-decimal-place discrepancy between the two solutions had been grossly magnified by the iterative process inherent in solving the equations. He was left with a picture of two vastly different weather systems.

Lorenz later told *Discover* magazine, "I knew right then that if the real atmosphere behaved like this [mathematical model], long-range weather forecasting was impossible."

Lorenz had immediately realized that it was the combination of nonlinearity and iteration that had magnified the microscopic three-decimal-place difference in the two computer runs. That the results were so far apart means that complex nonlinear dynamical systems such as the weather must be so incredibly sensitive that the smallest detail can affect them. As the new aphorism goes, the effect of a butterfly flapping its wings in Hong Kong can create a rainstorm in New York. Suddenly Lorenz and other scientists became aware that in deterministic (causal) dynamical systems, the potential for generating chaos (unpredictability) crouches in every detail.

At first it may seem unfair, or at least an exaggeration, to call a weather system chaotic just because we can't predict it. If our ability to predict is faulty isn't that because we just lack all the necessary detail or we don't have the right equation? The answer is no. What Lorenz had seen was that because of the iterated nature of nonlinear equations (which represent the interconnected nature of dynamical systems), no amount of additional detail will help perfect the prediction.

To understand why this is the case, let's run through a little demonstration of what can happen in iterations. The demonstration involves some strings of numbers, but don't worry, this is not higher math. What we're really interested in is following the patterns that will be quickly apparent.

Doubling a number is very simple. Recall again the first equation (for exponential growth) that Alice found on her blackboard.

Figure 4.2

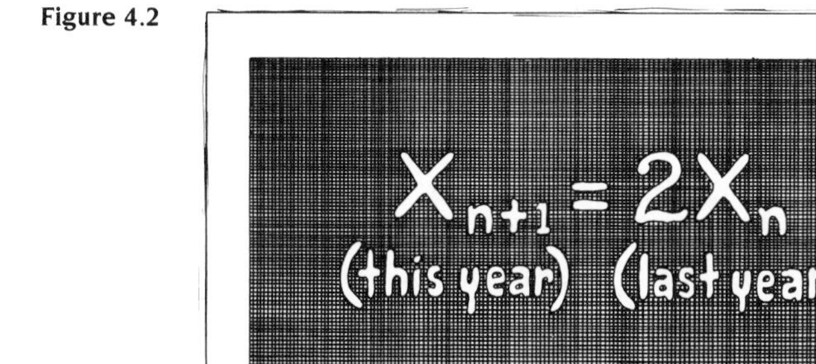

The equation says that this year's crop is twice last year's. If X_1, the first year's crop, is 1, then next year's crop, X_{n+1} will be 2. The equation generates the sequence for the years that follow: 2, 4, 8, 16, 32, 64 . . . (The three dots indicate that the sequence goes on forever.)

Or starting with $X_1 = 1.5$ we get the sequence for the following years: 3, 6, 12, 24, 48 . . .

Up to this point everything is straightforward. But now let's employ another one of those mathematician's tricks to enable us to generate some long strings of numbers and compare them as we wish. The trick is this: Continue doubling the number but knock off the integer part and keep the decimals. For instance, if X_1 (the first year) = 0.9567, then $2X_n$ (otherwise known as X_2, the next year) = 1.9134. Employing the mathematician's stratagem, we now drop the integer so that X_2 = 0.9134.

Let's see what sort of a series we get by starting with X_1 = 0.5986. . . . The series is: 0.1972 . . . , 0.3944 . . . , 0.7888 . . . , 0.5776 . . . , 0.1552 . . . , 0.3104 . . . , 0.6208 . . . , 0.2416 . . . , 0.4832 . . . , 0.9664 . . . , 0.9328 . . . , 0.8656 . . . , 0.7312 . . . , 0.4624 . . . , 0.9248 . . . , 0.8496. . . .

This appears to be a random sequence of numbers as if the iteration is leading to chaos. But let's look at this phenomenon in more detail.

If it happens that X_1 contains an initially simple order in the way its decimal digits repeat, then a correspondingly simple pattern will always be found during iteration. For example, if X_1 = 0.707070, the iteration generates the pattern 0.414141, 0.828282, 0.656565, 0.313131, 0.626262, 0.252525, 0.505050, 0.010101, 0.020202, 0.040404, 0.080808, 0.161616, 0.323232, 0.646464, 0.292929, 0.585858, 0.707070 . . .

After seventeen iterations we're back to the original number again; the cycle will repeat itself again and again.

Choosing a number with a more complicated pattern will create an even longer cycle before the string begins to repeat. But, provided that the starting numbers are rational, the pattern will eventually curl back. Rational numbers, remember, are those that can be expressed in terms of a ratio of integers as 1/2, 2/3, 3/4 and therefore always have a finite decimal form (1/2 = 0.5, ;1/4 = 0.25) or an infinite = 01010101). When rational numbers are fed into this simple number-doubling iteration they always generate ordered patterns.

But what about irrational numbers, which can never be written down as a ratio of integers? Their decimal expression contains no order; each digit appears at random. Mathematicians have verified that an irrational number like pi can be calculated to many millions of decimals without any repetition occurring. It seems ironic that pi, the number used to calculate the circumference of what many consider the most perfect and ordered object of our imagination—the circle—can never be calculated exactly. Even in the Euclidean world, order and chaos go hand in hand.

What happens when an irrational number is used as the initial input in our number-doubling sequence? The result is an infinite string of numbers containing no apparent order. Each new number occurs at random. Chaos appears to blossom from the very irrationality enfolded in the original number. In fact, the simple exponential growth equation, or number-doubling equation, is one way of producing strings of random numbers in a computer. Chaos and chance could be thought of as actually being *unfolded* out of the infinite complexity contained in the original irrational number.

A striking property of iterative equations is their extreme sensitivity to initial conditions. If X_1 in the number-doubling equation is changed very slightly, then the sequence will soon diverge from the original. It was precisely this property that was discovered by Lorenz in his weather calculations. In the nineteenth century scientists had always assumed that a small error in initial data would either be averaged out, or would, at most, produce a small effect. But where iterations are concerned, small errors can be rapidly amplified.

Think of that rational number 0.707070. What happens if we make a slight error in the fourth decimal place, an error of 1/10 of a percent, and write 0.707170?

At the first iteration the error is minor. Instead of the 0.414141 we got in the original sequence, the new series starts 0.414341. The second iteration swells the error more distinctly. Instead of 0.828282, the new second term is 0.828682. For the rest of the sequence, instead of the original 0.656565, 0.313131, 0.626262, 0.252525, 0.505050, 0.010101, 0.020202, 0.040404, 0.080808, we now have 0.657365, 0.314731, 0.629462, 0.258924, 0.517849, 0.035698, 0.071396, 0.142792, 0.285584. By the eleventh iteration the slight error has ballooned so much that the new sequence has completely diverged from the original. The original series repeated itself after seventeen numbers. The new series does not have this pattern.

The iteration reveals the extreme sensitivity of the equation to its initial conditions, its initial numbers. This sensitivity applies equally to rational and irrational numbers when they are iterated in nonlinear equations.

But it isn't just numbers that behave this way. Scientists observe the same dynamics in fluids. The final destination of a small eddy of blood in the bloodstream is exceptionally sensitive to its initial position. Neighboring points in the blood may continue to flow side by side, may oscillate around each other or end up in completely different parts of the fluid. Even our own aging can be thought of as a process in which the constant iteration of our cells eventually introduces a folding and divergence that changes our initial conditions and makes us slowly fall apart—drawn to death by perhaps the ultimate strange attractor.

In the physical world, different systems display different degrees of sensitivity to the iterations they undergo. One design of a plane wing produces a rapid magnification of the fluctuation budding around ice crystals on the wing surface, a magnification which occurs so quickly it may create a turbulence that will cause a plane crash. Other wing designs, however, are impervious to the same icing conditions. As we saw with period doubling, iteration at one rate produces stability, but when the rate is pushed past certain values, the system reels toward chaos. Though, as Feigenbaum discovered, the scale of critical values is the same for many systems, each system suffers its own nonlinear conditions where iterations will begin to flail out of control.

The movement of the type of nonlinear iteration found in so many systems can be visualized in terms of a baker kneading dough to make bread. With his fists the baker stretches out the dough and folds it over on itself, repeating this stretching and folding over again and again. In fact, mathematicians call what happens to a nonlinear equation when it is iterated "the baker transformation." This transformation has the effect of moving neighboring points in the dough away from each other. A series of elastic threads placed in the dough would eventually become stretched and folded into a very complicated, unpredictable (and hence chaotic) pattern. Mathematically, this process of stretching and folding takes the form of a strange attractor.

The baker transformation governs the growth equation. The Verhulst formula is guided by the dynamism of two opposing effects, one the stretching factor (X_n), and the other a folding back ($1 - X_n$). In this way the output of the previous iteration becomes the input for the next.

N = 0

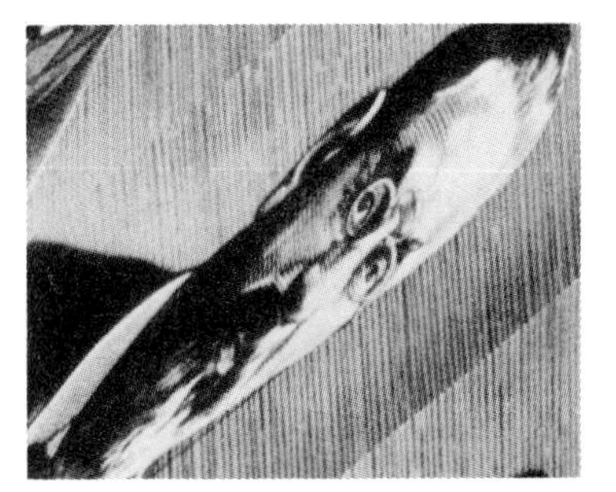

N = 1

N = 10

N = 48

N = 241

Figure 4.3. The stretching and folding of change is illustrated by this computer mosaic of the father of modern chaos, Henri Poincaré. Poincaré's image was digitalized by physicist James Crutchfield so it could be stretched mathematically as if it were painted on a sheet of rubber. Crutchfield uses the image to show how positive feedback or iteration can transform things. Iterating the computer formula, Poincaré's image is stretched on the sheet diagonally, and the leftover part is reinserted on the other side. The number above each panel tells how many iterations of this process have taken place. As the iterations go on Poincaré's face is scrambled randomly until it's completely homogenized.

However, as the folding operation continues, it may happen that some of the points come close enough to their initial positions for the image to reappear. In other words, a brief intermittency of order occurs before the iterative folding shears the points apart again. Crutchfield's equation makes a momentary return to a state close to the initial conditions (known to scientists as the "Poincaré recurrence") much more likely than it would be in typical chaotic transformations. In "typical" chaos, the chances of Poincaré's face showing up again as the system iterates are astronomically small, particularly if there is any background interference. A small spike in the electrical signal to the computer, for example, would become folded in by the iterations and would destroy the original information.

STRETCHING IT

Equations like the growth equation with Verhulst's added nonlinear term are guaranteed to generate a totally chaotic sequence with complete determinism, that is, you can determine all the terms going into the equation. Nevertheless, the calculations that follow from this iteration are something of a hoax because they're performed on a computer or even worse, a pocket calculator. As it turns out, this fact tells us something significant about chaos.

Computers generally carry their calculations out to sixteen decimal places. So with each simple operation there's always some rounding off involved. For instance, if the number 5 occurs in the sixteenth decimal place, it may have been because the sixteenth and seventeenth places were . . . .49 or . . . .51. The uncertainty about the actual value of the digit in the sixteenth place is generally so small that it never worries anybody. A pocket calculator only works to eight decimal places and how often do you need that last place?

But in the iterative equations of change, where the results of each stage of the calculation are fed back into the next (representing the feedback that exists in real systems such as fluid flows), the initial uncertainty about the sixteenth decimal place begins to accumulate and distort the results of each iteration. In fact, after fifty rounds of stretching and folding on the procrustean bed of iterations, the uncertainty is so serious that it swamps the calculation. Although the iterations are deterministic, the round-off error exploits the limitations of the computer and makes any prediction meaningless.

But suppose we use a bigger computer and allow for more decimal places. Suppose we build a computer that is as large as the universe and capable of carrying out calculations to thirty-one places.

Even with the round-off error minimized to the order of one part in 10^{31}, determinism and predictability sputter. For after only 100 iterations of this universe-large computer, our infinitesimal error will have gutted the calculation. Given the speed with which normal computers do iterations, predictability vanishes within a fraction of a second when highly nonlinear equations are concerned.

Figure 4.4. Video chaos. Aiming a camera at its own monitor produces endless feedback and the shape of chaos.

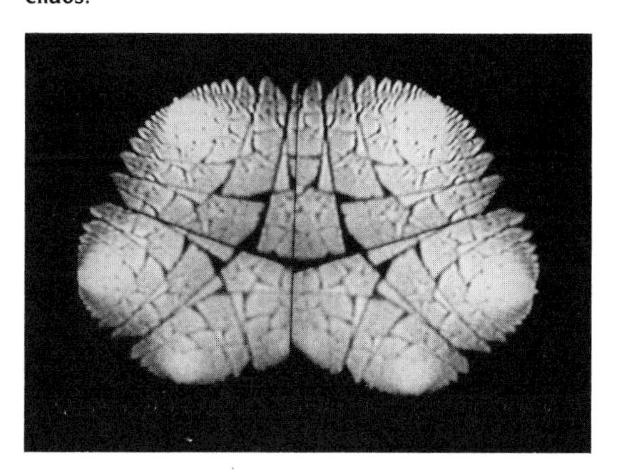

Chaos physicist Crutchfield says, "The consequence of measuring with only finite precision is that the measurements are just not good enough: chaos takes them and blows them up in your face." The butterfly effect. The scientists of change wax poetic over the issue of sensitivity to initial conditions.

Across the actual universe, the sensitivity of tiny numbers to iteration is mirrored by the fact that in the buzz of planets and orbiting electrons, correlations are constantly being swamped by an accumulation of microscopic changes.

In one scientific article Crutchfield, J. Doyne Farmer, Normon H. Packard, and Robert Shaw, four of the pioneers of chaos, explain that the sensitivity of dynamical physical systems is so great that perfect prediction of the effect of a cue ball striking a rack of billiard balls is impossible. "For how long could a player with perfect control over his or her stroke predict the cue ball's trajectory? If the player ignored an effect even as minuscule as the gravitational attraction of an electron at the edge of the galaxy, the prediction would become wrong after one minute!"

Why? Because the equations governing the hard billiard balls have an iterative nonlinearity, so that the movement of the system defined by the equation is infinitely sensitive to the changing movement of everything else—the air pressure, temperature, the nap of the table, the muscle tone of the billiards player, his or her psychology, the flight of neutrinos from a supernova millions of light-years away, the gravity of an electron. Iterating the nonlinear equation reveals this vast sensitivity to interconnectedness—which materializes in scientists' computers as unpredictability, chaos, the strange attractor.

This vast sensitivity suggests another slant on wholeness. Instead of thinking of the whole as the sum of all parts, think of it as what

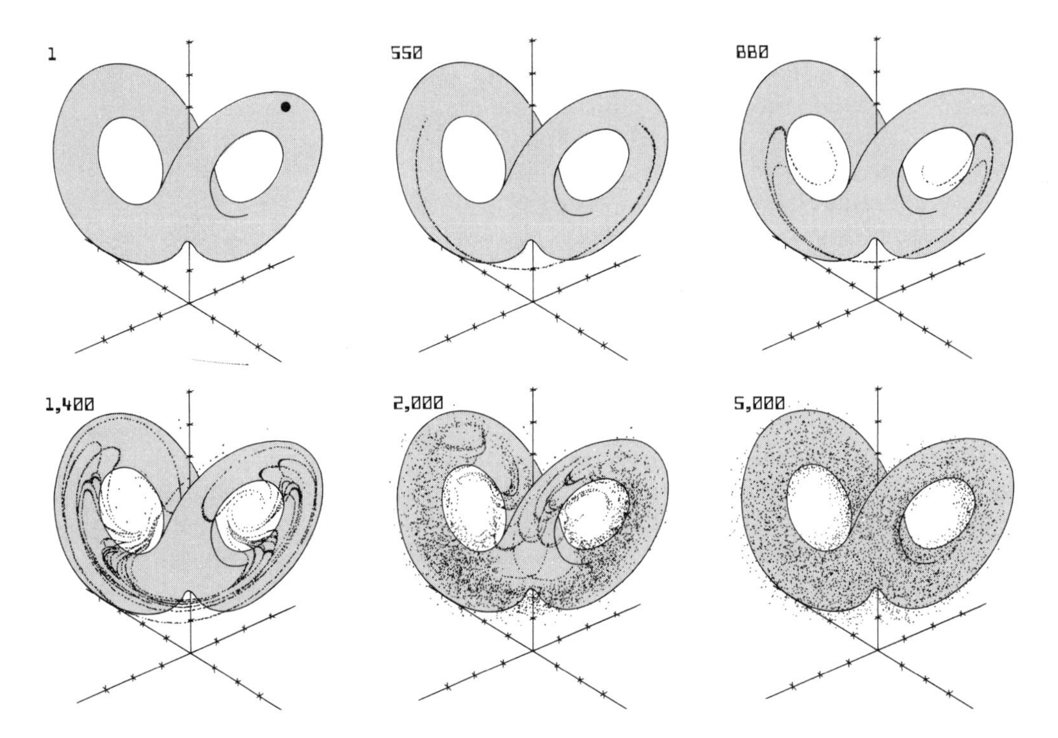

Figure 4.5. An illustration of how order leads to chaos. Start with a point in phase space. That point represents a vast complexity that underlies the system. In this case the system is the changing weather studied by Lorenz. The underlying strange attractor he discovered for weather systems has been named the Lorenz attractor.

Though the initial point of measurement (first panel) seems certain, it is in reality related by feedback processes to everything else in the system and so has an immense uncertainty built in. As the system iterates (that is, as its "parts" feed back into each other) the complexity and uncertainty begins to reveal itself. The point in phase space where the initial measurement was taken begins to stretch and fold into a cloud of uncertainty that takes the shape of the strange attractor. Very quickly the equation shows that the true state of the system (the weather outside) could be anywhere on the attractor. Chaotic systems such as the weather are said to be locally unpredictable but globally stable. Global stability means that they always take the shape of their strange attractor.

The strange attractor is not only the shape of unpredictability, it's also the shape of the weather's dynamical qualities and a picture of its interaction with the whole.

rushes in under the guise of chaos whenever scientists try to separate and measure dynamical systems as if they were composed of parts. It is the round-off error, what physicist Joseph Ford calls the "missing information" that spills in at the seventeenth or thirty-first or 5-millionth iteration and obliterates prediction. The missing information (the whole) is "implicated" in dynamical systems by a thin infinite thread of diminishing decimal points in the equations modeling dynamical processes. Through this thread as through the neck of a balloon, the whole is pumped by iteration until it explodes the equation.

Los Alamos theoretical physicist Frank Harlow says the uncertainties or errors, the missing information, in knowing the initial conditions of dynamical systems are similar to the "seeds" which produce turbulence and chaos: the butterfly's wings, a rough spot of ice crystals on the surface of a plane wing, an electron on the outskirts of the galaxy. Anything might be a seed if it's in the right place in the right dynamic. Iteration pumps microscopic fluctuations up to a macroscopic scale.

On a philosophical level, chaos theory may hold comfort for anyone who feels his or her place in the cosmos is inconsequential. Inconsequential things can have a huge effect in a nonlinear universe.

In fact, cosmologists speculate that if the initial conditions at the big bang had varied by as much as a single quantum of energy (the smallest known thing we can measure), the universe would be a vastly different place. The whole shape of things depends upon the minutest part. The part *is* the whole in this respect, for through the action of any part, the whole in the form of chaos or transformative change may manifest. That transformative "part," the incipient whole, is the "missing information" which through iteration traces out the system's unpredictability. The shape it traces is the strange attractor.

Figure 4.6. Scientists are discovering many different kinds of strange or chaotic attractors. The one below is called the Rössler attractor, named after the theoretical chemist Otto Rössler who got the idea for it by watching a mechanical taffy puller repeatedly stretching taffy out and folding it back on itself. Rössler imagined what would happen to two raisins in the taffy and wrote down the equation that would describe their divergence. Rössler's attractor has been observed as the shape of growing turbulence in fluid flows and chemical reactions. Nearby points in the system are stretched around this shape again and again, creating folds within folds. Quickly the points separate, and the numerous folds make it impossible to say just where the points are on the attractor. The attractor is the shape created in phase space by the "missing information," the shape of uncertainty. Are attractors shapes through which the infinitely complex order of the whole reveals itself?

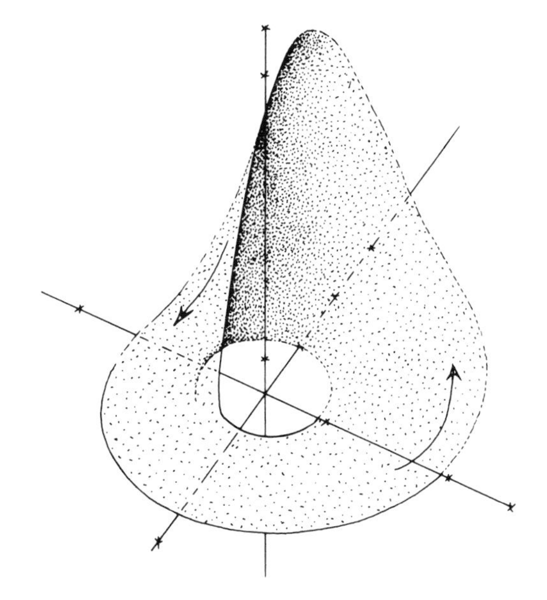

A number of physicists believe there is a connection between the principle of "missing information" in chaotic systems and Gödel's famous incompleteness theorem. In the 1930s Kurt Gödel stunned the mathematics community by showing that important logical systems like arithmetic and algebra will always contain statements that are true but which cannot be derived from a fixed set of axioms. There will always be missing information, a hole (or should we say *whole*) at the center of these logics, Gödel found.

His proof of the incompleteness theorem was based on the paradox of the liar. Instead of the Cretan saying "all Cretans are liars," Gödel proved a mathematical statement that said, "This statement is unprovable."

Gregory Chaitin, a mathematician at IBM's research center in Yorktown Heights, NY, uses a new information theory proof of Gödel's theorem to argue that what Gödel found was not just a mathematical curiosity. The iterative paradox, the (w)hole in the center of our own logics, the potential chaos of the missing information, applies naturally to many if not most of the things we think about, Chaitin believes.

Quantum mechanics in the early part of this century discovered in such laws as the uncertainty principle, complementarity, and the wave-particle duality that there are built-in limits to what we can observe about events at the microscopic level. Bohr postulated that at that level there is an unbroken wholeness not separable into parts or events. Thus it seems that twentieth-century scientists, from Bohr to Gödel to chaos theorists, are revisiting an ancient insight. In the third century B.C., Aristotle stated it in his *Nichomachian Ethics*: "It is the mark of an educated mind to rest satisfied with the degree of precision that the nature of the subject admits, and not to seek exactness when only an approximation is possible."

Quantum mechanics is a revolutionary theory because it sees the microworld as basically statistical and indeterminate, not "exact." Chaos theory comes from classical physics, from Newtonian cause-and-effect determinism —reductionism—which is still thought to govern the large-scale world. Most scientists had thought that at least here in a world of traffic patterns and rain clouds, cause and effect must dominate. Even if we can't learn to predict and control such things perfectly, it was believed, we can approach closer and closer to the ideal. But in the mirror of determinism, we have glimpsed an indeterminate invasion. The scientists of chaos have discovered that determinist systems which maintain themselves by oscillation, iteration, feedback, limit cycles (systems including most everything of interest to us) are vulnerable to chaos and face an indeterminate (unpredictable) fate if pushed beyond critical boundaries.

Two pots of soup heated on a stove under exactly the same conditions will behave differently. Conditions for dynamical systems are never identical, but for the most part we can ignore differences with impunity because they don't become magnified, turning the familiar into the chaotic. We have traditionally appreciated the simple regularity of order in our familiar world, neglecting the infinitely higher orders (or chaos) woven within it.

But phenomena such as a heart attack reveal that inside localized order, inside regular iterations and oscillating systems, strange attractors lurk. In fact, it has recently been found that the *normal* heartbeat is irregular and follows a subtle strange attractor.

Our very life and health depend upon living within layers of order and disorder. Physician Paul Rapp has pointed out that chaos theory offers the possibility of treating "convulsive disorders" like epilepsy by "resetting parameters" so that the brain's oscillations come back inside the normal chaotic boundaries and the convulsions stop.

Richard Day, a University of Southern California economics professor, has shown that many of the important equations in economics are subject to the kind of iteration that leads to chaos and undermines predictability. Day says that economists usually assume that external shocks and unexpected events upset economic cycles. But he has found that the cycles themselves are inherently chaotic. "Periods of erratic cycling can be interspersed with periods of more or less stable growth. Evidently the 'future' behavior of a model solution cannot be anticipated from its patterns in the 'past.'" And what happens to the models is just what happens in reality: Regular order is interspersed with chaotic order.

Evidently familiar order and chaotic order are laminated like bands of intermittency. Wandering into certain bands, a system is extruded and bent back on itself as it iterates, dragged toward disintegration, transformation, and chaos. Inside other bands, systems cycle dynamically, maintaining their shapes for long periods of time. But eventually all orderly systems will feel the wild, seductive pull of the strange chaotic attractor.

Fittingly, it was Poincaré who first noticed the sensitivity of iterated systems to their initial conditions. An avid gambler, the great French mathematician observed that the subtle differences in the flick a croupier gives to a ball in a roulette wheel can make an immense difference for which slot the ball finally drops into. The croupier's cry can now be understood as the cry of chaos, of order, of change and as the sonorous cry of the whole: "Round and round it goes, where it stops, nobody knows."

THE MIRROR

ORDER TO *Chaos* TO ORDER

A. A *violent order is disorder: and*
B. A *great disorder is an order.*
These two things are one.

WALLACE STEVENS
"CONNOISSEUR OF CHAOS"

A *mirror whose world we can enter and whose inhabitants can enter into our world is like a portal with two sides.* We *have just explored the landscape on one side of that portal, a landscape which includes turbulence, period doubling iteration, and strange attractors.* On *this side we've observed the ways in which orderly systems grind into chaos, and seen signs that what we have called the* Yellow Emperor's *spell is being broken.* That *breaking spell is scientific reductionism, the belief that the universe is fundamentally made up of parts.* By *shattering this spell, scientists have discovered that efflorescing all around them is a new kind of magic.* As *we'll soon see, it's a magic that comes from the other side of the mirror, from beyond its portal, in the landscape where chaos gives birth to order.*

Be*fore we can enter the landscape on that side of the portal we must pass through the mirror.* Here, *held for a moment on its turbulent surface, in the very frame of the mirror, we are at the fertile boundary line between order and chaos.* It *is a place of strange beauty, so we will pause for a while to look around and savor the odd experience of being on both sides of the mirror-world at once.*

CHAPTER 0

The Book of the Yellow Emperor *says:*
"*But unborn it is not basically* Unborn, *shapeless it is not basically the* Shapeless."

LIEH-TZU

MEASURES OF CHANGE

Period-doubling plots, twists in phase space, the Lorenz, Rössler, and other strange attractors. These conceptual devices that we encountered on the "order to chaos" side of the mirror are like x-rays, giving scientists the ability to glimpse the evolving skeleton of nonlinear change. The vivid images spawned by these devices have been powerful dispellers of the reductionist idea—primarily because they have provided scientists with a new way to measure things. In fact, they are instances of a revolution that is taking place in scientific measurement. For hundreds of years reductionism—or the idea that the world is an assemblage of parts—has been supported by powerful mathematical techniques which quantify reality. By quantifying reality, parts can be added and subtracted. Since scientists using mathematics of quantification have been immensely successful in making discoveries and predictions, scientists' belief in reductionism has grown.

But, as we've seen, when scientists study complex systems, the notion of parts begins to break down so that quantification of such systems becomes impossible. So scientists wanting to study dynamical systems have turned to another approach to measurement —*qualitative* mathematics. In the old quantitative mathematics the measurement of a system focuses on plotting how the quantity of one part of the system affects the quantities of other parts. By contrast, in qualitative measurement, plots show the shape of the system's movement as a whole. In a qualitative mode, scientists don't ask, How much of this part affects that part? Instead they ask, What does the whole look like as it moves and changes? How does one whole system compare to another?

In this chapter—representing our pause in the portal which is the exact center of the turbulent mirror—we'll examine several kinds of qualitative measure beyond the ones we've already seen, and we'll see how qualitative measurement has helped to catapult scientists into a new perspective on reality from which they have gained quite stunning views of how order, chaos, change, and wholeness are all woven together.

RUBBER MATH

In the past three decades, nonlinear change has yielded up many secrets to topology, a branch of mathematics which deals with the way shapes can be pulled and distorted in a space that behaves like rubber. In topology straight lines can be bent into curves, circles pinched into triangles or pulled out as squares. However, topologically not everything is changeable. Intersections of lines, for example, remain intersections. In mathematicians' language an intersection is "invariant," it can't be destroyed no matter how much the lines are twisted. The number of holes through an object is also invariant in topology, meaning that a ball may be transformed topologically into a pancake or a cube, but never a donut.

Back in the 1960s, at the beginnings of chaos theory, mathematician Stephen Smale realized that topology could be used to visualize dynamical systems. By bending, twisting, and folding a topological shape, it's possible to represent how a system moves. By topologically transforming one shape into another, it's possible to compare very different dynamical systems.

Smale decided to topologically investigate a period-doubling system discovered in 1927 by a Danish engineer. Balthasar van der Pol had used an electrical feedback loop to translate an oscillating electrical current into tones of the same frequency on a telephone. Inexplicably, Van der Pol found that when he increased the current in his electrical loop, the tones jumped to smaller and smaller multiples of the frequency. Between the jumps were spurts of noise, chaos. Van der Pol was encountering intermittency through the iterations of feedback. Unfortunately he didn't understand the implications of what he'd heard and simply dismissed the "noise" (which was in fact created by the conflicting attraction between the higher and lower frequency) as a "subsidiary phenomenon."

Smale decided to model the van der Pol oscillator topologically. Instead of trying to follow the trajectory of this complex dynamical system around in phase space, Smale envisioned phase space itself as stretching and folding as the system moved in the boundary area between higher and lower frequency attractors. The result is called Smale's horseshoe.

Imagine a rectangle squeezed and stretched out into a bar. Bend the bar into a horseshoe and embed that in a rectangle. Then stretch, squeeze, and fold this rectangle into another horseshoe and repeat the whole process again and again. This is what happens to a system that is period doubling its way to chaos, Smale realized.

French mathematician René Thom used another sort of topological fold to describe nonlinear change where systems undergo abrupt, discontinuous transitions from one state to another.

Thom studied systems driven into sudden and radical change not so much by the pure oscillations of their own innards as by external forces. The sudden transformation of a corn kernel in a popper, the collapse of a support beam in a bridge weighted down by one pound too much load, the dramatic translation of water into ice at 0° C or into steam at 100° C, the flicking on or off of a light switch—are all examples of what Thom calls "catastrophes."

Thom's insight was that all such abrupt changes can be classified topologically as one of seven "elementary catastrophes." Each ca-

FIGURE 0.1

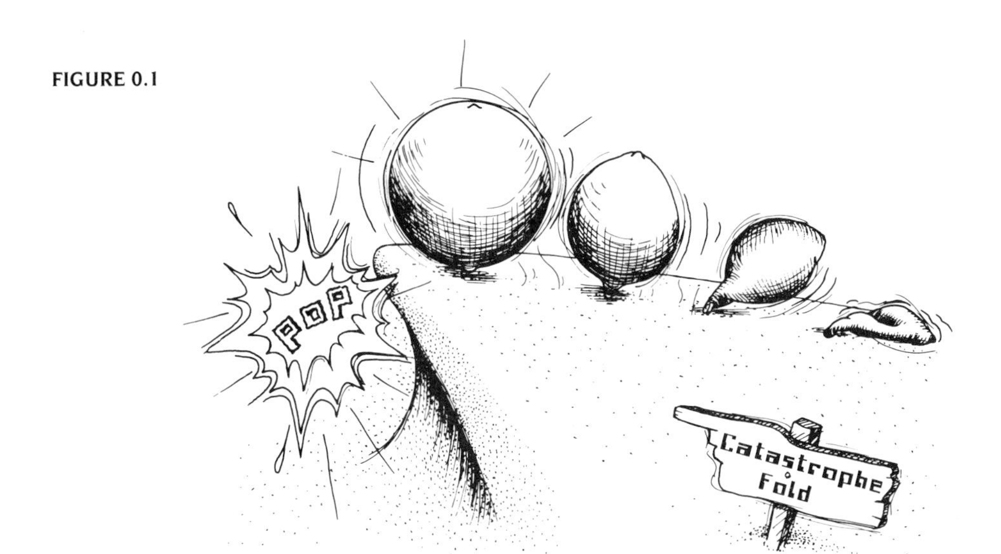

tastrophe involves folds in the phase space across which the system moves. The folds are created by the system's "control variables," that is, the external elements that push the system's behavior.

Thom's first form of catastrophe is simply called the "fold."

Consider a balloon being blown up by someone at a party (FIGURE 0.1). The control variable in this change is air pressure, because reducing or increasing air pressure alters the balloon's dynamics.

As the air pressure in the balloon increases, the system approaches the edge of the catastrophe fold. If pushed too far it falls over the fold—into oblivion. After passing through its "fold catastrophe," the balloon bursts and the system no longer exists.

Though the fold catastrophe is by far the simplest in Thom's catalogue of seven universal catastrophes, it's a description which can be applied to such complicated phenomena as a rainbow, a shock wave, and a supersonic aircraft. Any system dominated by a single factor or control variable can be portrayed on this topological "map."

When the number of controls is increased from one to two, a second catastrophe "map" comes into play. Now we have a system that can be pushed in two different ways. Consequently, the topological "map" of what Thom calls the "cusp catastrophe" has two dimensions, which can be represented by a sheet of paper deformed so that a fold appears. The control variables or important influences on the system can be pictured as pushing the system around on the folded surface of the paper.

Take, for example, the behavior of your pet dog. The biologist Konrad Lorenz argued that the dominating factors in a dog's behavior, in other words its control variables, are rage and fear. Using Thom's cusp catastrophe fold, it's possible to visualize how rage and fear could act to suddenly transform a dog's behavior.

Suppose that your dog is approached by another dog. At first your pet is filled with rage at the sight of this interloper and commences to yap and bark and growl menacingly. This state is indicated by the right-hand side of the drawing above. But what happens if the ap-

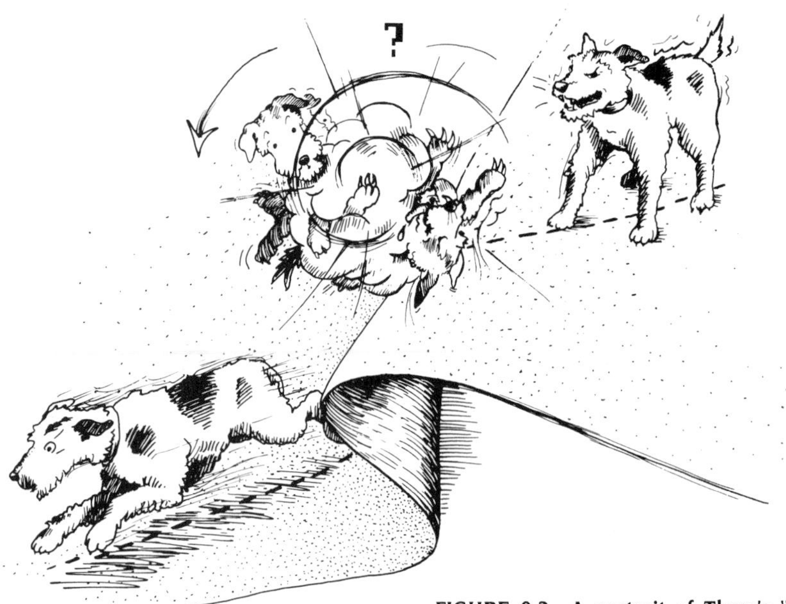

FIGURE 0.2. A portrait of Thom's "cusp catastrophe" representing the internal state of a dog as it moves from rage to fear. When the dog reaches the edge of the fold, it enters a twilight zone and could go back into its fight behavior or fall over the fold into a totally new behavior, flight.

proaching dog is much bigger than your pet? Now your pet begins to experience a little fear, and his "behavior point" is pushed to the left. Nevertheless, he is still on the upper region of the catastrophe fold, the region signifying aggressive behavior. So as far as an onlooker is concerned, nothing has changed. Your pet continues to bark and snarl.

Then as your pet's fear mounts, its behavior point will move closer and closer to the catastrophe fold, though your dog still barks and barks.

Finally, however, he reaches the very edge of the fold. Here the smallest change in one of the control variables (rage and fear) could send him over the edge. Let that big interloper dog take just one more step and your pet tumbles into a mental "twilight zone" where he leaves the upper surface of the behavior space altogether and reappears at the bottom of the fold in a totally new behavior—flight.

Thom's topological study dramatically illustrates how a small change of rage or fear within a dog's mind will generally produce only an imperceptible difference in behavior, but at a critical point it may produce a very abrupt behavior change.

In place of the dog turning tail, we could substitute the crash of the stock market or the response of a heavily loaded beam in a bridge. Thom's catastrophe theorem shows that whenever a system can be described using a single behavior variable influenced by two control variables, that is, two major influences, then it can be represented by the cusp catastrophe of FIGURE 0.2. This catastrophe fold works as a description of manic-depression, the breaking of sea waves, prison riots, lasers, the flow of polymers, crystal symmetries, or of decision-making processes. The nonlinear systems described by Thom's catastrophe theory are stable for most of their lives. It's only when they venture to the edge of one of these catastrophe folds that they suffer discontinuous change. Indeed, the point attractors and limit cycles discussed earlier can be included in Thom's catastrophes, but this time sketched out on a phase space which can be deformed topologically. Thom's catastrophe portraits give insights into how such apparently stable systems can suddenly transform. Thom's treatment of nonlinearity brought an important ingredient to the turbulent science. Nonlinear dynamical systems, whether chaotic or stable, are so complex they are unpredictable in their detail, indivisible in their parts—the smallest influence can cause explosive change (for example, the smallest nuance can push the dog from aggression to flight or flight to attack). Nevertheless, Thom found a way to represent such systems as a whole, using the qualitative measure of topological folds.

A MATTER OF DEGREE

The ability to compare the nonlinear changes taking place in vastly different systems is the great appeal of Thom's theory. This is also the appeal of the qualitative measure called the Lyapunov number, named after the Russian scientist who invented it. The Lyapunov measure makes it possible for clouds, electrical activity in the brain, and the turbulence of rivers to be compared on the basis of their *degrees* of order and disorder.

Picture a superhighway with several lanes of traffic. In the middle of the day, cars slip along in a steady stream, with no bunching up or great gaps in the flow. Adjacent lanes move at different speeds but the difference isn't very great. A truck traveling at 55 miles per hour is slowly overtaken by a car doing 60. Like the smooth flow in a river, a characteristic of this movement is that neighboring elements (in this case automobiles) either stay together or depart from each other only gradually.

Now picture rush hour. The increased flow of cars creates chaotic conditions analogous to turbulence. Cars accelerate and switch lanes. Some bunch up, others speed into vacant stretches of highway. Neighboring cars may separate rapidly with one accelerating far ahead into a free lane and the other trapped in a long crawling line.

The Lyapunov number is a measure of how fast neighboring points in a river or on a highway or in any dynamical system separate from each other. Consequently, it measures how quickly correlations in the system are broken down and how rapidly the effects of a small perturbation can spread.

A similar measure describes how the system's "information" changes. For example, the relative positions of all the cars on the highway could be fed into a computer and monitored from minute to minute. This information defines the overall traffic flow. If the flow is regular, the cars in each lane keep almost the same relative distance from each other and the information scarcely changes or changes in a simple regular way. But during rush hour the information changes wildly. Scientists talk about the original information becoming "lost," though it might be more accurate to think of it as transformed.

An analogy to this loss, or transformation, of information is passing a message in English through a cipher machine that jumbles it up into apparently meaningless letters or digits. In one sense the meaning of the message is lost; in another it has simply been transformed because a reverse or deciphering transformation could restore it completely. However, the transformations of information can become so subtle and complex that reversing the process is impossible.

A MEASUREMENT EXPERIMENT: A STRANGE TALE

There is no question that scientists engaged in the tricky problems of trying to measure change in the turbulent mirror have frequently stumbled upon some outlandish things. The following case in point illustrates that there is a universe of surprisingly subtle order awaiting revelation by a holistic approach to measure.

Four researchers at the University of California at Santa Cruz decided upon an ingenious method for gauging the degree of order in a devilishly simple chaotic system many of us have around the house: a dripping faucet.

How is such a system chaotic? In a turbulent river, each element of flow, each small "part" acts as a contingency for every other part. The river generates its contingencies out of its wholeness. Water under certain pressures leaking from a tap also generates contingencies. Thus the four scientists reasoned that by measuring one "part" or aspect of the dripping tap water they could obtain a snapshot of the whole system. And, by constructing a phase space out of their measurements, they would try to see whether the system was under the influence of a strange attractor, and perhaps even get a picture of the attractor.

To carry out their experiment, the researchers placed a microphone under a faucet (FIGURE 0.3), which leaked "like an infinitely defective drummer," and plotted the time intervals between successive drops, a measure of the degree of chaos. They chose this aspect of the system to measure, though they might also have measured how long it took for the drops to form on the faucet or the relative weight of the drops.

On a graph the researchers noted the different intervals of over 4,000 drops. The result was surprising. Certainly it would be logical to expect that plotting something purely random would produce a random, blobby shape. But in fact that's not what happened. Moment by moment, as the scientists recorded the time interval between drops, the graph points jumped around chaotically. Nevertheless, as more and more points were laid down on the graph, a shape emerged from the mist that looked remarkably like the cross section of a strange attractor known as a Hénon attractor—that is, an attractor generated by iterating an equation according to rules first established by Michel Hénon of the Nice Observatory in

FIGURE 0.3

France. Later, when the four scientists increased the water pressure in the faucet a little, they found eerie and experimentally reproducible shapes that appeared to be cross-sections of other "hitherto unseen chaotic attractors."

The Hénon attractor invites comparisons with a ring system around some science fiction planet. But its truly fantastic quality is revealed by zooming (computing) in for a look at more detail in one of these rings. Like the structure of gaps and debris in the real rings of Saturn (see Figure 1.21), inside the ring structure of the Hénon attractor another ring structure appears, similar to the one at the larger scale. In turn, if one of these fine rings is explored at a higher magnification, more rings unfold.

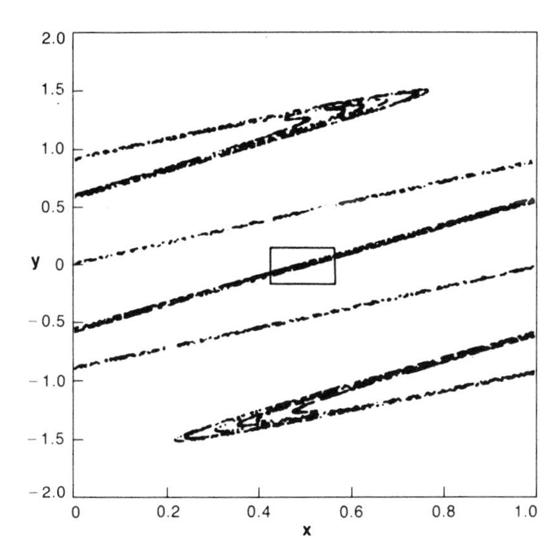

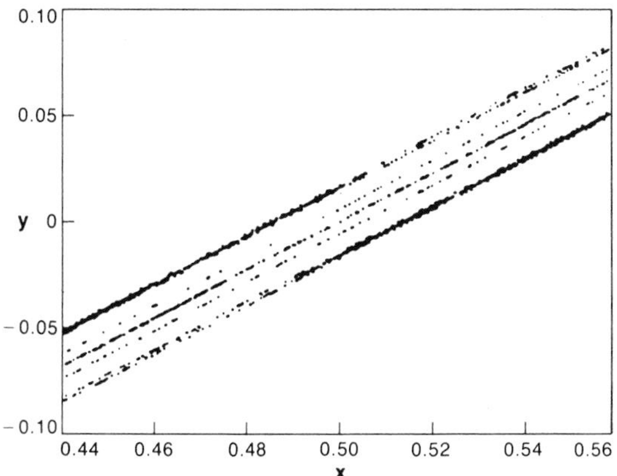

FIGURE 0.4. The Hénon attractor, similar to that found by plotting the random drips of a faucet. Zooming in on a small section of the attractor reveals its self-similarity.

This giddy world of the infinitesimal recalls an advertisement for a brand of brown sauce popular in England during the 1940s. The picture showed Daddy bringing a bottle of sauce to the table. On the label of the bottle was a picture of Daddy bringing a bottle of sauce to the table, and on the label of that bottle was a picture. . . .

David Ruelle has proposed that the Hénon attractor, Rössler attractor, Lorenz attractor—strange attractors of all kinds—are really like Chinese boxes of subtle order. This wild attracting order exists in the cracks of things, inhibits a fractional realm that lies between the first, second, and third dimensions of the familiar world with its point attractors, limit cycles, and well-run tori. As we're about to see, probing this curious fractional realm requires yet another form of qualitative measure.

THE FABULOUS FRACTAL

Smale, Thom, Lyapunov, Ruelle, and others have created important qualitative instruments for seeing the movement of order, chaos, and change in the nonlinear world. But more than anyone else, one mathematician has revolutionized the turbulent science with his discovery of a qualitative measure that has immortalized the mirror-world's intricate beauty. His discovery has also revealed that the mirror-world is uncannily the same as the world we inhabit every day.

Benoit Mandelbrot's education was irregular and his mind stubbornly visual. He says that when he sat for the crucial entrance exams at France's prestigious École Polytechnique he was unable to do the algebra very well but succeeded in getting the top grade by translating the questions mentally into pictures.

Even today Mandelbrot claims not to know the alphabet so that using a telephone book is an ordeal, but he can *see* things that other people can't. For example, he says, "I do not program computers myself, but have found ways of working very interactively with several outstanding people: students and assistants, but also colleagues like Richard F. Voss. As a matter of fact, I developed a skill for helping 'debug' programs that I cannot read, by analyzing the wrong pictures these programs produce."

Frustrated by the highly abstract mathematics being taught him in school, the young Mandelbrot cultivated a fascination for the geometric (or rather, *non*geometric) irregularity in the world around him. He was driven by a sense that he later expressed in an aphorism that, he says, "has gained the supreme accolade of becoming an instant cliché." His driving geometric intuition was that "clouds are not spheres, mountains are not cones, coastlines are not circles and bark is not smooth, nor does lightning travel in a straight line."

After his years at school, Mandelbrot's career became as irregular as the shapes he was interested in. He studied aeronautics at the California Institute of Technology, was sponsored at the Institute for Advanced Study in Princeton by the brilliant mathematician John von Neumann, and did research in a number of fields. "Every so often I was seized by the sudden urge to drop a field right in the middle of writing a paper, and to grab a new research interest in a field about which I knew nothing. I followed my instincts, but could not account for them until much much later."

In 1958 Mandelbrot became a research staff member and in 1974 a Fellow at IBM's prestigious Thomas J. Watson Research Center in Yorktown Heights, New York. There, in a smoothly curving glass structure set into the hills of Westchester County, his intuitions began to coalesce. A new geometry emerged in his mind, unlike anything that had gone before. Mandelbrot conceived the fractal.

The name comes from the Latin *fractua*, which means irregular, but Mandelbrot also liked the word's connotations of fractional and fragmented.

In the early flush of his idea, he used fractals to plot stock prices and produced mathematical forgeries that were good enough to fool experts in the field. His fractals showed that major recessions mimic both monthly and daily price fluctuation, so that the market is self-similar from its largest to its smallest scales.

Turning to the problem of noise in data transmission, Mandelbrot created a workable model out of his new geometry; and, using no astronomical data, he mathematically visualized a distribution of galaxies in the universe that astrophysicists have since confirmed. "I became very aware that self-similarity, far from being a mild and uninteresting property, was a very powerful way of generating shape." By "self-similarity" Mandelbrot means a repetition of detail at descending scales—the repetition of the picture of "Daddy's sauce."

Though Mandelbrot is a tireless missionary for his fractals, these days it's hardly necessary. The great theoretical physicist John Wheeler has said that in the past people could not consider themselves scientifically educated unless they understood entropy. In the future, Wheeler insists, "no one will be considered scientifically literate who isn't equally familiar with fractals."

Wheeler's statement speaks to the fact that in the past twenty years Mandelbrot has done an impressive job in conveying his vision. It is now clear that fractals embrace not only the realms of chaos and noise but a wide variety of natural forms which the geometry that has been studied for the last two and a half thousand years has been powerless to describe—forms such as coastlines, trees, mountains, galaxies, clouds, polymers, rivers, weather patterns, brains, lungs, and blood supplies. Just as physics had tried to lump together a vast range of subtle properties of nature under the general heading of "chaos" or "disorder," these most exquisite forms in nature, with all their rich detail, were ignored by conventional geometry. Consider the way the turbulence of wind and water gouges out and sculpts the starkly dreamy shapes of canyons, mesas, and undersea grottos. Do such places lack order? Mandelbrot avers that Euclidean geometry is "dull." In revenge, he has shown that irregularity is exciting and that it is not just noise distorting Euclidean forms. In fact this "noise" is the bold signature of nature's creative forces.

Take, for example, the blood supply in our bodies. In an anatomy textbook, veins and arteries with their repeated branching may appear chaotic, but looked at in more detail, it becomes clear that the same complex branching is repeated for smaller and smaller blood vessels right down to the capillaries. The same is true of a mountain. Seen from forty miles away the mountain's outline is quite recognizable, yet at the same time it's irregular. The closer we drive, the more detail is present and even when we begin to climb the mountain we notice the same pattern of irregularity and detail in the individual rocks. The complex systems of nature seem to preserve the look of their detail at finer and finer scales. This question of scale comes up again when we look at the marvelous shapes and structures of nature in a book of photographs taken through microscopes and telescopes. Images from vastly different scales evoke a feeling of similarity and recognition.

But how could something that measures thousands of light-years across have anything in common with objects that can be encompassed in the hand or on the head of a pin? Could it be that similar mathematical laws or principles of growth and form are operating at such different scales? If this is true, Mandelbrot realized, then these laws must have little to do with classical geometry, where scale is a notion so obvious that it is of little or no importance. Could one create a measure of irregularity that was based on scales?

Mandelbrot's first step in examining the question of scale and concretizing his vision of an irregular yet orderly world was to turn to some curiosities and anomalies of mathematics that had surfaced toward the end of the nineteenth century and been dismissed by mathematicians. Could it be that these mathematical oddities held important clues to the complexity of nature?

In 1872 a mathematician named Karl Weierstrass precipitated a minor crisis in mathematics when he described a curve that could not be mathematically "differentiated." The ability to differentiate, that is, to calculate the slope of a curve from point to point is a central feature of calculus. Calculus was invented independently by Newton and Leibniz some 200 years before Weierstrass. Newton's new laws of mechanics dealt with regular change and with rates of change, and he needed a mathematics to describe various forms of gradual change; he found it in calculus.

The idea of slope is a fairly intuitive one. You experience it every time you climb a hill. Slope is really the same thing as a gradient. In the case of a railway track, the value of the gradient can sometimes be seen written on a pointer as, say, 1:200. This means that for every 200 feet of track the altitude increases by 1 foot. The slope or gradient of a road may be even higher; in mountain areas, a side road may have a gradient as high as 1:6 or 1:5.

Of course, individual roads are not perfectly regular and tend to dip and rise, so the gradient printed on a map or shown on a road sign is an average value. With more accurate surveying, it's possible to determine the gradient in increasingly smaller intervals and take into account individual variations of the road. Newton's calculus went one step better. The mathematical equation for the climbing road determines the slope or gradient at each point. This determination is mathematically equivalent to differentiating the equation of the curve.

Ever since Newton, mathematicians have been contentedly differentiating curves and functions and their slope. There were always problems, however, if the curve was discontinuous, that is, if the road suddenly disappeared and appeared somewhere further on. How could you have a slope right at the edge where the road finished? But leaving aside those special cases, all curves, mathematicians believed, must have slopes. In more formal language, they believed that a continuous curve can always be differentiated.

Newton's calculus seemed secure until, at the end of the nineteenth century, a mathematician named Debois Reymond stepped in and presented Weierstrass's equation for a curve that was continuous but so complicated that it could never have a differential.

The result was a panic among mathematicians that took some fifty years to resolve. In the end they were forced to concede that such anomalous curves could exist. But mathematicians also consoled themselves with the thought that any curve so complex and absurd must have absolutely nothing to do with the real world.

Another bombshell burst around 1890 when Giuseppe Peano discovered what was called a "space-filling curve." A curve is nothing more than a line that bends and deforms and, as every schoolchild knows, a line is one dimensional. Mathematicians took it as a matter of common sense that any curve, no matter how much it bends, must be one dimensional. A plane (a piece of paper for example) is two dimensional. The plane and the curve are perfectly distinct in terms of their dimensions.

Nevertheless, Peano had constructed a curve that twisted in such a complex way that it actually filled the whole plane of the paper it was drawn on. There was no point on the plane that Peano's curving line would not include. This created an unpleasant situation for mathematicians. The very two dimensionality of the plane lay in its set of points. What did it mean if all these points were also on a one-dimensional line? How could an object be one dimensional and also two dimensional?

Nicolai Yakovlevich Vilenkin in *Stories About Sets* expresses the reaction of mathematicians: "Everything had come unstrung! It is difficult

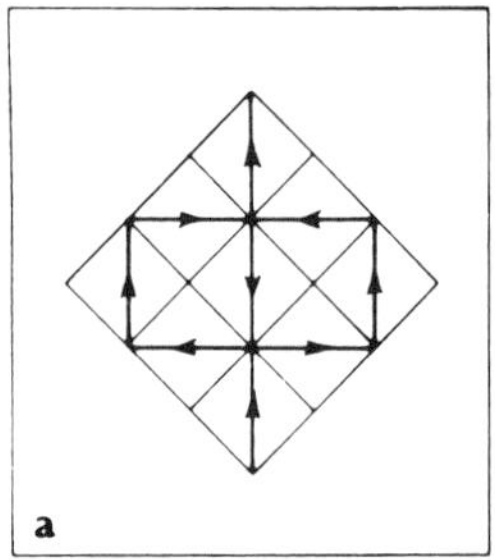

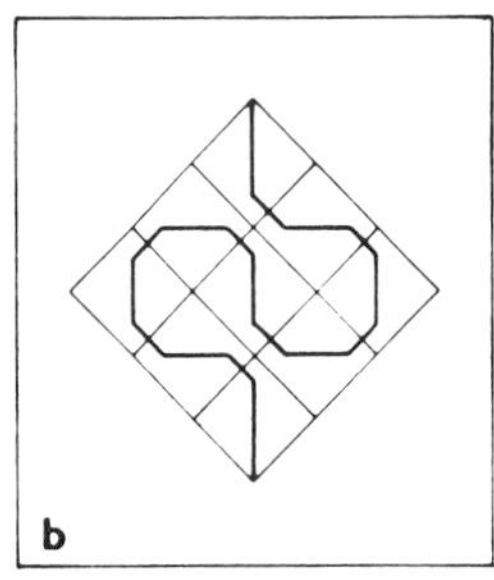

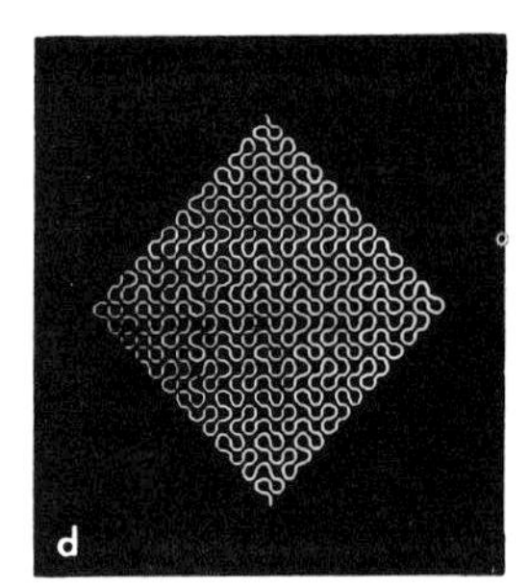

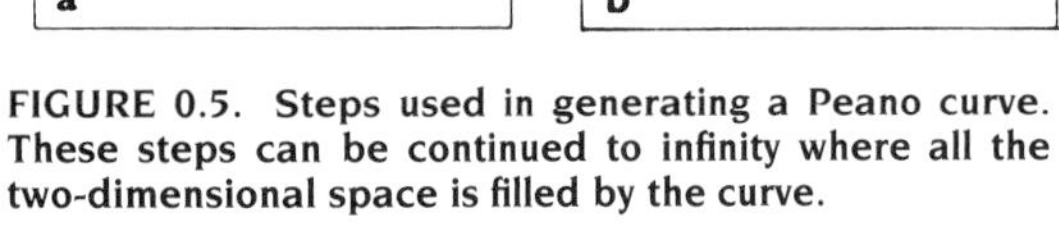
FIGURE 0.5. Steps used in generating a Peano curve. These steps can be continued to infinity where all the two-dimensional space is filled by the curve.

to put into words the effect that Peano's result had on the mathematical world. It seemed that everything was in ruins, that all the basic mathematical concepts had lost their meaning."

These outrageous curves with no slope and with an ambiguity of dimensions were enormously upsetting. Mathematicians' only hope was to dismiss such things as a mere chimera of abstract thought, a mathematician's joke posing no threat to the ordered way in which mathematics and geometry described nature. The great Poincaré himself adopted such a defensive stance. He called the strange curves "a gallery of monsters."

Seventy years after Peano, however, Mandelbrot took such curves seriously and by following their implications was able to turn the tables on mathematics. He showed convincingly it was not the case that the monster curves have little to do with the geometry of the world. Quite the reverse. In them, he demonstrated, lies the secret of the way to measure the irregularity of the real world. The secret of fractals.

What exactly is a fractal and how is one made? FIGURE 0.6 shows the generation of a fractal that has its origin in the "snowflake" curve constructed by Helge von Koch in 1904. Essentially, the "Koch island," or snowflake, is created through a process of iteration in which each step is taken on a smaller scale. In this way a curve of considerable complexity is produced, containing an extraordinarily high degree of detail.

With their many bays, inlets, and promontories, Koch islands are reminiscent of real islands—except that they're far too regular. True islands require more sophisticated fractals to describe them. But, at the least, Koch islands show a degree of complexity quite foreign to conventional geometry. Clearly this simple fractal points to something very new about the way mathematics can be used to describe the forms of nature.

FIGURE 0.6. Repeated application of the generator to the sides of a triangle (an initiator) creates a jagged snowflake in which the triangle is repeated on smaller and smaller scales.

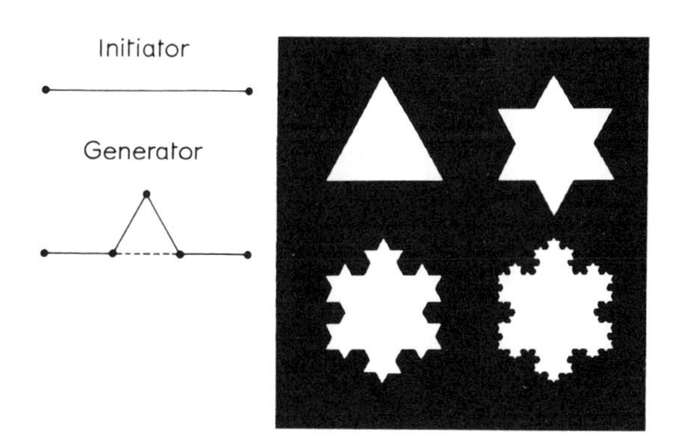

For a mathematician this figure holds less obvious surprises as well. The first comes when an attempt is made to measure the island's perimeter, that is, to find the length of its coastline.

In fact this can be put in terms of a real-world question: How long is the coastline of Great Britain? This was exactly the question Mandelbrot posed in a new classic paper. The answer won him renown.

Countries, of course, want to know the length of their coastlines and boundaries. When a boundary is drawn between two countries, say between Canada and the United States or France and Spain, it's a good idea if both parties agree on the length. At first sight this seems to be a straightforward problem with a straightforward solution—just measure it. But in fact gazettes and geographical texts give different mileage for the same coastline or boundary. How can this be? Is it a matter of negligent surveying? Of poor calculation?

We might think that the question of the length of Britain's coastline could be settled by taking a good map and running a piece of thread all around the coast, then reading the result from the scale printed at the bottom of the map. But a moment's reflection reveals that the map tends to smooth out and omit fine detail. It only gives the broad turns of the coastline and leaves out the many bays and inlets.

The answer then must be to take a more detailed map. In that case the thread will be curved and looped around more detail. But this means that the coastline's length will be greater. Can this result be improved on? If a surveyor makes an accurate survey at, say, 100-meter intervals along the coast, it will be even more finely detailed. In turn the coastline will have a greater length.

But why stop here? Why not survey at 50-meter intervals—10 meters even? In each instance, finer and finer detail will be included and the thread will curve in more and more complex ways. By now it's evident that the more detail that is included, the longer the coastline gets. What if *all* the detail is included—rocks, pebbles, dust, even molecules? The true coastline must be infinite. Indeed the coastline of Britain is the same length as that of Manhattan or the whole of the Americas. They are all infinite.

This was the shocking conclusion Mandelbrot reached. But how can it be true? A little thought convinces us that any figure containing detail at progressively diminishing scales must have an infinite length. So clearly, what applies to the coastline of Britain also applies to the length of a Koch curve, to all fractal curves.

In practice we can agree on a conventional scale and ignore all detail below 100 meters or some other figure. This is equivalent to viewing a coastline "out of focus" so that details smaller than 100 meters across are smeared out. If they could agree on a scale, cartographers could measure and compare coastlines. However, from a mathematician's point of view such a compromise leaves much to be desired.

Since mathematically all coastlines with real detail must have infinite length, can such figures be compared at all? More surprises, for the answer Mandelbrot discovered is yes. However, the answer shifts the question from one about measuring the length quantitatively to a new kind of qualitative scale-based measure—the fractal dimension.

In order to understand fractal dimensions we need to shake up our commonsense ideas of what a dimension means. Most people believe that they have a pretty clear idea of this concept. Space is three dimensional. A wall or tabletop or piece of paper is two dimensional. A line or curve or edge is one dimensional. And finally, a point or even a set of points is of zero dimension.

The dimensions we meet in everyday life are straightforward: 0, 1, 2, or 3. But are things really that simple? What, for example, is the dimension of a ball of string?

From far away the ball looks like a point and therefore has zero dimension. But from a few feet away everything is back to normal and the ball is three dimensional. But what happens if we come very close? We see an individual thread that is twisted up and wrapped around. The ball is composed of a twisted line, and is therefore one dimensional. Even closer this line turns into a column of finite thickness, and the thread becomes three dimensional. Closer still, we lose the thread in favor of the individual fine hairs which twist around and around each other to make up the thread—the ball has become one dimensional again.

In other words, the "effective dimension" of the ball keeps changing from three to one and back again. Its apparent dimension depends on how close we get to the ball. So we see that dimension is not necessarily as straightforward as it first appears. Maybe all dimensions in nature aren't any clearer than this; they depend on the way we look at them.

Mandelbrot has gone so far as to say he thinks that when his fractal geometry highlights the inextricable relationship between object and observer it is in keeping with the other great scientific discoveries in this century, relativity and quantum theory, which also found an interdependence between observer and observed. The quantitative measure—on which science has been based—is also challenged by this insight. The length of the coastline depends on what quantity we choose as the measure. If in the end quantity is a relative concept—it always involves some smearing out of details—then it is considerably less precise than we believed. In place of a quantity, such as length, Mandelbrot puts the qualitative measure of effective fractal dimensions, a measure of the relative degree of complexity of an object.

Disconcerting as it may at first seem to admit that objects in nature have such "effective dimensions," the concept makes it possible to work out a fractal dimension for a coastline and to discover that this is a fractional number greater than one. If a curve or a coastline's fractal dimension is close to one, the coast is very smooth and has no fine detail. The greater the number is above one, the more irregular or chaotic the coastline is, with this irregularity persisting at smaller and smaller scales.

How are irregularity and detail connected to fractal dimension? Imagine scattering grains of rice uniformly across a map. There may be, say, 10,000 grains and this collection could be said to characterize the two dimensionality of the map. A straight line drawn across the page passes through only 200 grains, so only 2 percent of the grains lie on the line. The vast majority lie in other regions of the plane. But suppose now the line twists and curves so that it passes through more and more grains of rice, reaching not only the grains of rice but even the individual points in the plane. As more and more of these points are crossed it becomes clear that the dimension of the line lies closer to that of a plane (two) than a line (one). In fact twisting fractal lines have dimensions that are fractional, such as 1.2618, 1.1291, 1.3652, and so on. The coastline of Britain has a fractional dimension of 1.26.

Now we can understand a little better the fractal curve created by Giuseppe Peano. This curve has become so extremely irregular at infinitely decreasing scales that its fractal dimension is two. Why two? Because Peano's line has so many twists to it that it reaches every point in the plane. However, despite its extreme self-contacting complexity, it never crosses itself.

In general fractals are characterized by infinite detail, infinite length, no slope or derivative, fractional dimension, self-similarity, and (as was done to produce the Koch coastline) they can be generated by iteration.

We can now understand why fractals and strange attractors are so intimately connected. Remember, in a phase space diagram, a strange attractor is traced by the point which represents the system. In its movement the system point folds and refolds in the phase space with infinite complexity. Thus, a strange attractor is a fractal curve. Fractal shapes have self-similarity at descending scales. For systems under the folding and stretching influence of the strange attractor, any single folding motion of the system represents (though in a unique instance) a mirror of the entire folding operation.

Wherever chaos, turbulence, and disorder are found, fractal geometry is at play.

But this suggests the rather astonishing conclusion that chaos and turbulence must be born out of the same underlying processes as mountains, clouds, and coastlines, or as the organic forms of nature such as lungs, nervous systems, and blood supplies. The complexity of an ever-branching human lung can now be understood as a mirror image of the chaotic motion of a fast-flowing river. Both emerge from a fractal order.

It has generally been assumed that complicated forms must be generated by means of a complicated process. For example, complexity in the human body is taken as a manifestation of very involved instructions for growth and development. But fractals are at one and the same time highly complex and particularly simple. They're complex by virtue of their infinite detail and unique mathematical properties (no two fractals are the same), yet they're simple because they can be generated through successive applications of simple iteration.

A FRACTAL SPACE VOYAGE

The realization that fractals are generated by simple iterations inevitably compelled Mandelbrot to try out his iterative geometry in the universe of pure mathematics. Mandelbrot says that in 1980 he was particularly inspired in this direction by reading some references in an old obituary of Poincaré about a peculiar problem the founder of nonlinear dynamics had once wrestled with. Mandelbrot puzzled over the same problem using his new fractal geometry. The result was like unearthing a diamond—only in this case the diamond was a stunning mathematical strange attractor.

Mandelbrot began by iterating a simple algebraic expression on a computer. This sent him on a voyage into the infinite two-dimensional sheet of numbers called the complex plane. The particular set of complex numbers Mandelbrot explored in this plane has since come to be named the "Mandelbrot set" and dubbed "the most complex object in mathematics." Mandelbrot remains enthusiastic about what he found.

"This set is an astonishing combination of utter simplicity and mind-boggling complication. At first sight it is a 'molecule' made of bonded 'atoms,' one shape like a cardioid and the other nearly circular. But a closer look discloses an infinity of smaller molecules shaped like the big one, and linked by what I proposed to call a 'devil's polymer.' Don't let me go on raving about this set's beauty."

Hundreds, perhaps thousands of computer adventurers have by now journeyed into the set using home computer variations of an iterative program explained by A. K. Dewdney in the pages of *Scientific American*. But explorers of the Mandelbrot set need have no fear of being imposed on by a crowd like tourists at the Grand Canyon. The unearthly Mandelbrot landscape—the mathematical strange attractor—is vast, in fact infinite, and "there are zillions of beautiful spots" to visit, says Cornell mathematician John H. Hubbard. He recommends, "Try the area with the real part between .26 and .27 and the imaginary part between 0 and .01."

Hubbard's fanciful-sounding invitation refers to coordinates on the complex number plane. Setting the figures on the equation is like setting dials on a spaceship and propels the iteration toward a coordinate formed by the intersection of two parts which are, for historical reasons, called "real" and "imaginary." Any complex number is made up of these two parts. And any complex number can be represented by a point in the complex plane. It's pretty much like locating Phoenix, Arizona, in an atlas map by finding the intersection of the letter J and the number 10. The main difference is that on the complex plane the number of possible intersections is infinite and the real and imaginary parts of the coordinates can be positive, negative, whole numbers, or decimal expansions.

The propulsion system that is jetting the computer into the Mandelbrot set is the equation $Z^2 + C$. Z is a complex number allowed to vary and C is a fixed complex number. The adventurer sets his or her two complex numbers into the equation and tells the computer to take the result of the addition of $Z^2 + C$ and substitute it the next time around (and the next time around after that . . .) for Z.

FIGURE 0.7

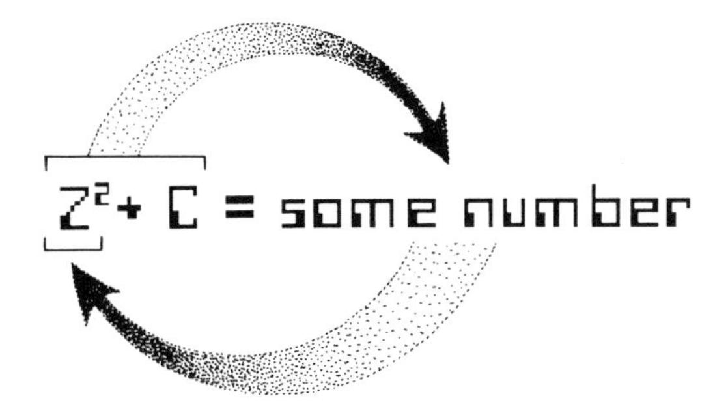

Thus the wild iterative spaceflight begins.

The computer whirs its way into the mathematical cosmos as the program searches for all complex numbers in the area that are not so large that they exceed the capacity of the computer to calculate them. The set itself consists of complex numbers C for which the size of $Z^2 + C$ remains finite no matter how many iterations the equation is put through.

On the computer screen, which records the number landscape you are entering, the Mandelbrot set first shows up as an ominous warty black object floating inside a circular bottomless well of points on the complex plane. The program guiding the journey takes a complex number and iterates it for up to a thousand turns. Does the number remain essentially the same through its iterations, approach infinity, or oscillate somewhere in between? The program is instructed to color or shade each point on the screen according to the answers it finds to these questions. In the black and white renditions that follow, the numbers that remain stable are the set itself and are colored black. The numbers that the iteration stretches toward infinity are colored a gray scale, with white indicating the numbers that iterated most rapidly to infinity. On the boundary of the set, the fate of iterated numbers is wild and uncanny.*

We can think of the boundary area as a terrain that lies between the finite solid world of the black inside of the set and the unstable limitlessness of the white and gray areas. This boundary is fractal.

In their book *The Beauty of Fractals*, Heinz-Otto Peitgen and Peter H. Richter describe this area in terms of a struggle. "A simple boundary between territories is seldom the result of this contest. Rather there is unending filigreed entanglement and unceasing bargaining for even the smallest areas." These smallest areas themselves have a bottomless depth because there is always an infinity of numbers between any two numbers on the complex plane. Consequently Mandelbrot adventurers can descend into the infinite well of $Z^2 + C$, examining the boundary in greater and greater detail, restricted in their flights of magnification only by the power (that is, the computing capacity) of their equipment.

The space you fly through is completely foreign but unsettlingly familiar, vivid yet totally abstract. The flight recorded in the next few pages was made by David Brooks, an engineer at Prime Computer Inc., Natick, Massachusetts, and "copiloted" by Dan Kalikow, another Prime engineer whose commentary accompanies the collection.

The names of the frames that log the trip are the ones assigned by Brooks or by other travelers who have previously explored these areas.

* The particular shade of gray used in the following illustrations indicates how far the numbers in this area are from the set and how long it takes for the computer to decide if the number is in the set, according to David Brooks, who wrote the computer program. Brooks says he reversed the natural order of grays so as to give a higher contrast with the black area of the stable numbers in the set.

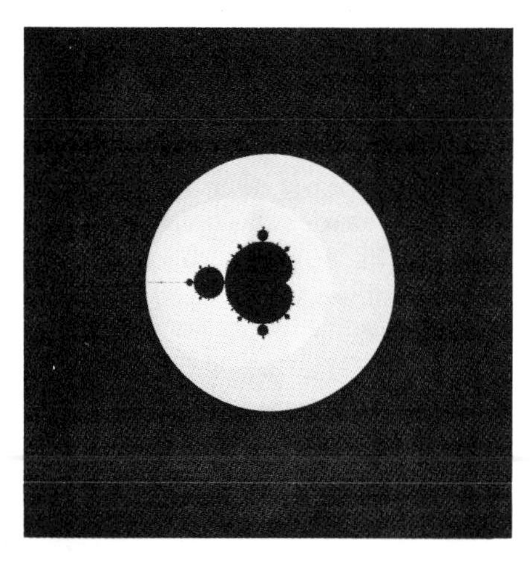

FIGURE 0.8. FRAME 1—GOD'S EYE VIEW
The journey starts high above the complex plane. Looming ahead is a well containing the Mandelbrot set. Like a fantastic planetary object, it's wreathed in envelopes of atmosphere made up of pools of complex numbers. The white numbers go to infinity when iterated—the pure whites go very fast, the grays less fast. The black ones lie solidly inside the set. At this "altitude" we can't see much detail.

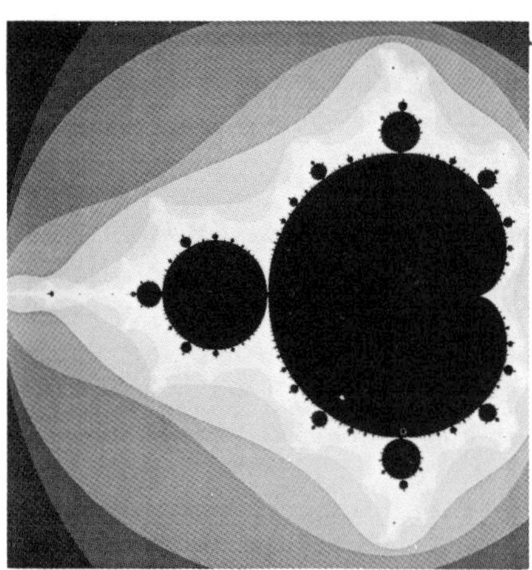

FIGURE 0.9. FRAME 2—MANDELBROT SET
Brooks has maneuvered his computer a little closer to the object, entering the outer reaches of the Mandelbrot atmosphere. As we head in we begin to see some detail of the atmospheric envelope that has been formed by the boundary between the set and the numbers surrounding it. What we'll be probing is the boundary line, which is fractal. Kalikow has set the "size" of Frame 2 as a reference point with which to compare deeper magnification. That means that all the subsequent frames are enlargements of this "natural size" object.

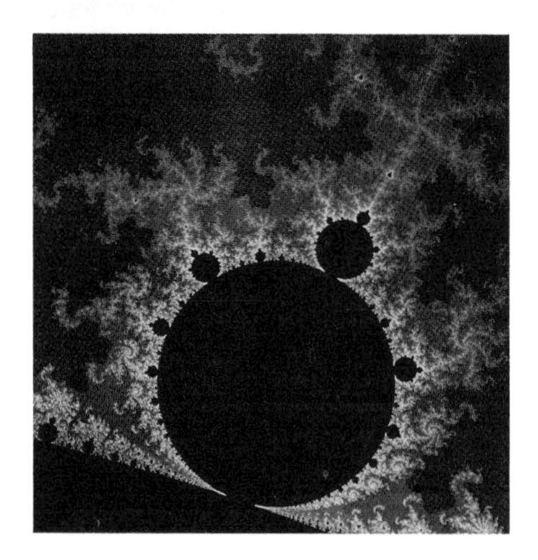

FIGURE 0.10. FRAME 3—COVER OF *SCIENTIFIC AMERICAN*
The Prime engineers zoom in on one of the buds. The picture here duplicates the *Scientific American* cover which made fractals famous. Even at this early magnification we're lost in the self-similarity of the Mandelbrot object. It's not self-sameness, however. Each bud and bud of a bud is slightly different. Note what Brooks calls a "mini-Mandelbrot" ascending like a space vehicle above the bud pointing toward the top right corner of the frame.

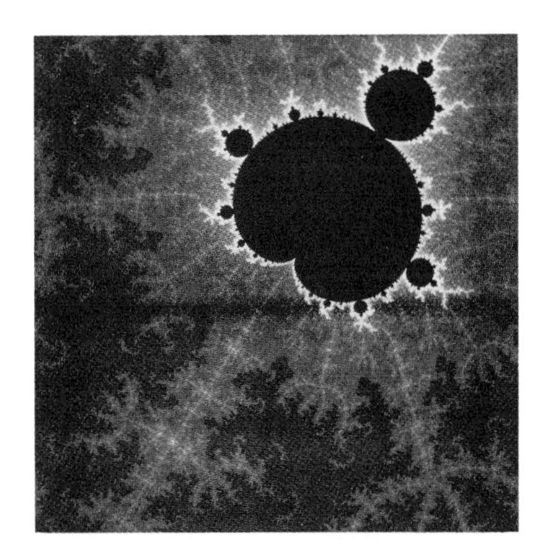

FIGURE 0.11. FRAME 4—MINI-MANDELBROT SET
We're descending now at dizzying speed. The magnification at this place in the journey is 2,500 times the reference size in Frame 2. John Hubbard and Adrian Douaday of the University of Paris have proved that the Mandelbrot set is connected, which means that all the miniature Mandelbrots are actually attached to the whole set by filaments. Kalikow says it occurred to Brooks that in Brooks's computer program he had "an ideal engine to test this assertion empirically. . . . So we aimed the microscope at a likely looking target." The target is the upside-down, heart-shaped cleavage at the bottom of the Mandelbrot object, an area called the "inflection point." Kalikow comments, "If there's a thread connecting this to the 'Mother Set,' it's got to be coming in here, we reasoned."

FIGURE 0.12. FRAME 5—FILAMENT
We're now up to 50,000 times the reference magnification. But no filament (which would be black like the set itself). "Nothing resolvable as a black line," Kalikow opines. "There *did* seem to be some little 'pearls' on some sort of string, but no string." Note the wave-crest structures rolling along the two "shores" of the mini-Mandelbrot. We'll be visiting a similar area later.

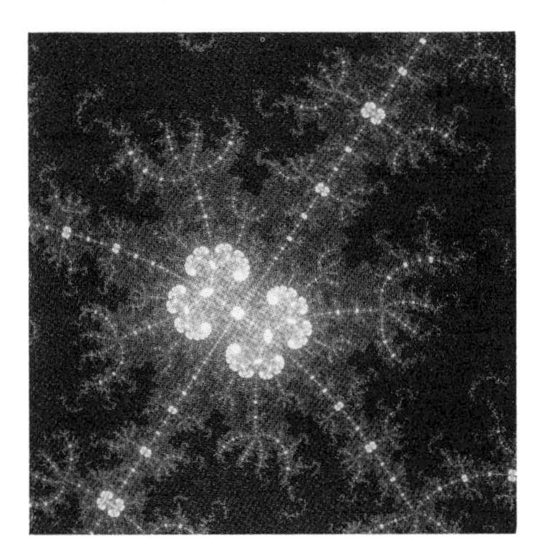

FIGURE 0.13. FRAME 6—PART OF FILAMENT
Darting down in our high-speed computer vehicle toward one of the "pearls" reveals a new level of detail, a filigree cluster. But still no black string. The magnification is 833,333 times the original size and took the computer seven hours to produce.

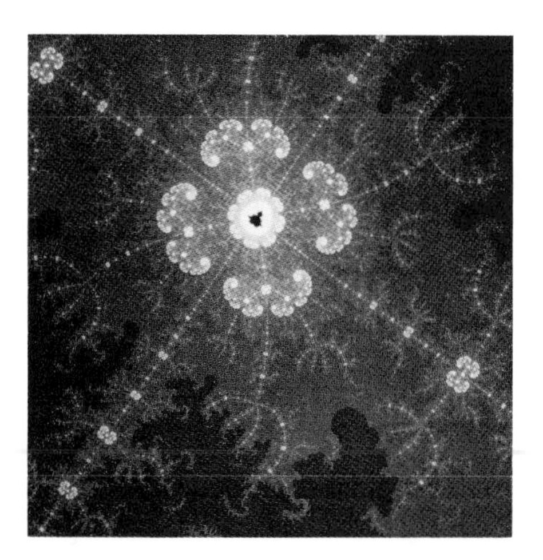

FIGURE 0.14. FRAME 7—PART OF A PART OF A FILAMENT
Again the filament vanishes as Brooks plunges after it toward the crossroads of the filigree cluster. However, brooding in the center, like an enigmatic scarab in the womb of thought, another micro-Mandelbrot has suddenly popped into the scanner.

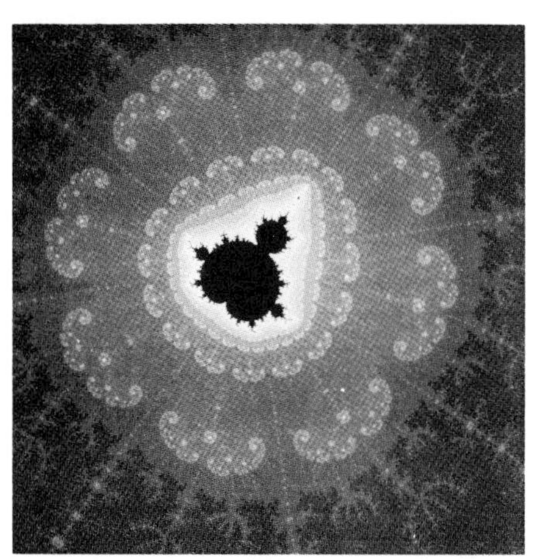

FIGURE 0.15. FRAME 8—ISLET
Hovering downward for a closer look, more pearls but no thread. Magnification 83,333,333. Kalikow says that it became clear that "no matter how small the steps we take in an attempt to land on one of the threads, we can never actually write one with a computer." Is it a parable? In the universe of the Mandelbrot object, the black of the unlocatable thread represents the finite, stable world which in the end, too, proves infinitely elusive. Is this an image of our inability to completely specify initial conditions?

But quickly we leave these deep thoughts behind and head back up toward the surface again, toward an area like the one we saw in Frame 5, to have a closer look at one of the two "shores" V-ing out from an "inflection point."

FIGURE 0.16. FRAME 9—SURF
Stately waves wheel out in curlicues and conch shapes spotted with mini-Mandelbrots. Another parable. Here float islands of order in a sea of chaos, worlds within worlds. Are we seeing how a simple iteration reveals the way a comprehensible order structures chaos? Or is it chaos that structures order? This is the turbulent mirror. In fact, the generation of Mandelbrot's mathematical set mirrors how real systems create and destroy the structures of our physical world.

FIGURE 0.17. FRAME 10—GORDIAN KNOT
We start a final breathtaking flight in Brooks's computer into the sinuous whirlstream of one of the "waves" in the previous frame. We're heading for an area around a "nano"-Mandelbrot.

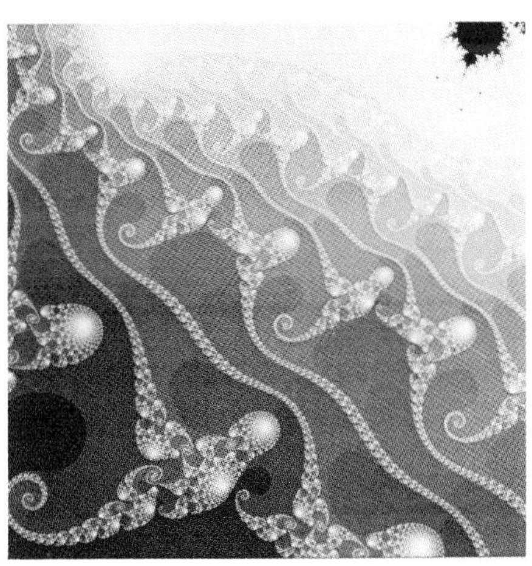

FIGURE 0.18. FRAME 11—THE N-FOLD WAY
The magnification at this depth is 2,702,702,702. In other words, if the reference Mandelbrot in Frame 2 were enlarged to the same scale as this picture, it would be a square 319,922 miles on a side, that is, a length one and one-third times the distance from the earth to the moon. Notice the nano-Mandelbrot in the upper right-hand corner, that strange attractor, floating like a persistent memory in a sea of white infinity.

FIGURES 0.19, 0.20. The connection between Mandelbrot's abstract mathematical set and the route to chaos can be demonstrated in the next two figures. FIGURE 0.19 is a diagram of a slice of the inside of the Mandelbrot set. FIGURE 0.20 on the next page is Robert May's period-doubling plot from Chapter 3.

As FIGURE 0.19 shows, the numbers inside the set are stable, that is, they don't change very much when they are plugged into the equation and the equation is iterated. But toward the boundary, the numbers period double. And those at the boundary of the set have period doubled to chaos.

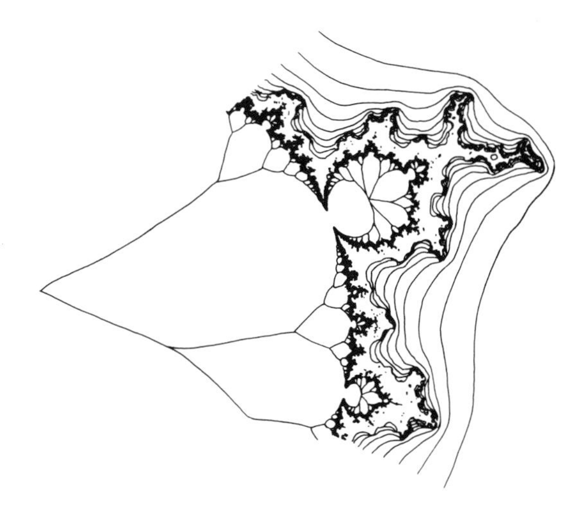

FIGURE 0.20. If the intermittency window in the period-doubling plot is magnified it reveals another period doubling inside. This window corresponds to those mini-Mandelbrots scattered on the complex plane. FIGURE 0.20's bifurcation points (where the attractors double) correspond to the Mandelbrot "buds" on the edge of the set.

Remember that this graph has been found to mirror the behavior of real systems such as insect populations. So there is clearly a connection between the fantastic mathematical world of the Mandelbrot set and the real world we live in.

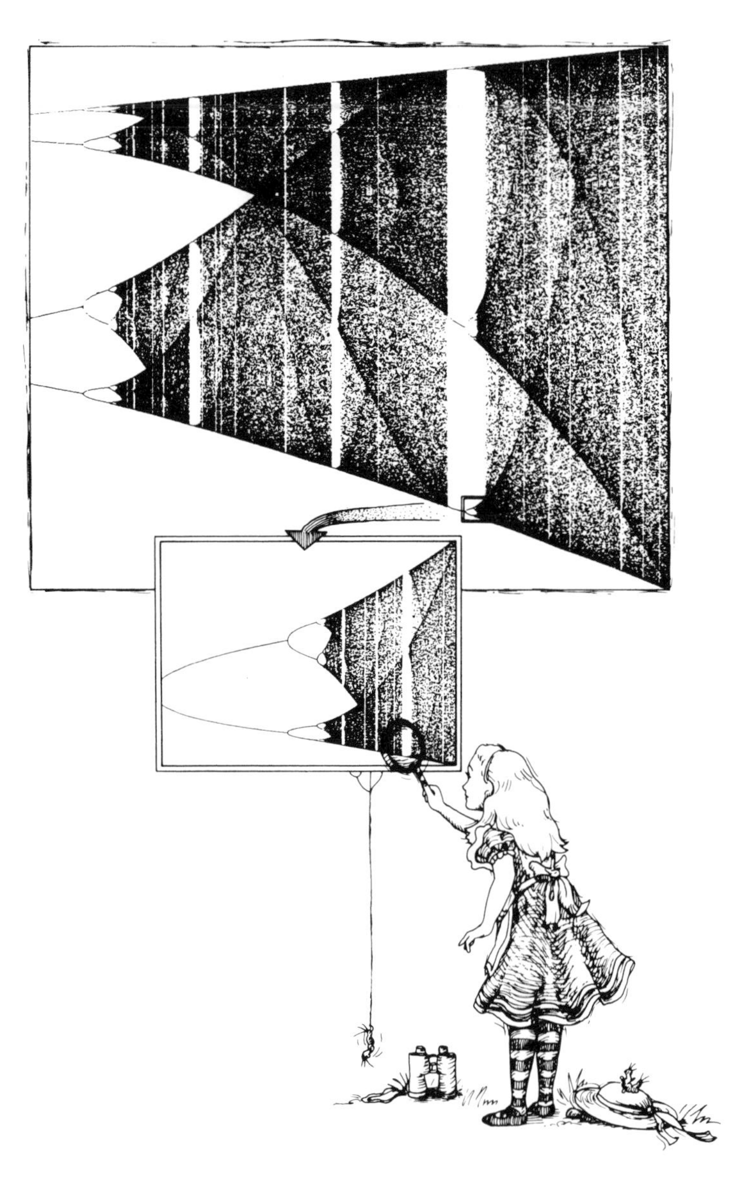

At the end of his journey, Kalikow sighs that the numerical 'reality' imaged in the Mandelbrot set is "far more abstract and eternal than mere physics. . . . It was always here . . . waiting to be seen. . . . Why do the shapes occur where they do? What's special about the numbers where the black shapes sit? How can they possibly be connected? They dust the complex plane like stars and galaxies clustering in ever-higher agglomerations, in an infinitude of shapes and levels."

The Mandelbrot set isn't the only fractal shape that can be generated by iterating certain equations of what Brooks calls "abstract and eternal mathematics." Many other equations have been found to possess a fractal nature. A centuries-old mathematical technique called Newton's method is also fractal. Newton's method enables you to find the roots of an algebraic equation by first guessing what the root is and then applying the method to the guess. The result is a number that comes closer to the root. The method is then applied to this number, and the iteration goes on in this way until you're satisfied that you've come as close to the root as necessary.

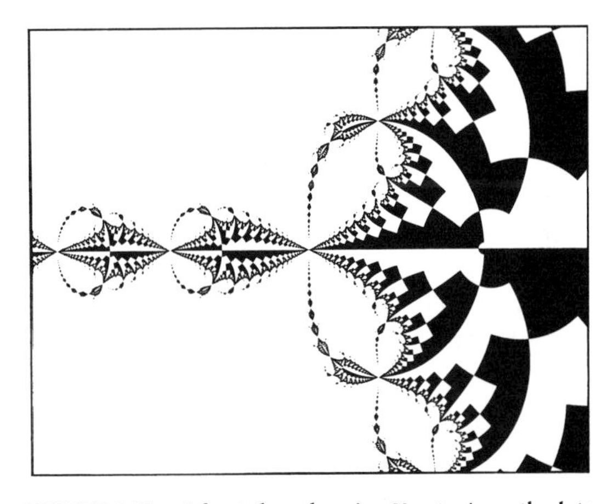

FIGURE 0.21. A fractal made using Newton's method, to find the cube root of − 1. One root lies in the white area. Two other roots are in the black area. The boundary region between the three roots is fractal, repeating the patterns at diminishing scales. Every point on the spiraling boundary touches upon the three areas containing the equation's roots.

Applying this technique on a computer will produce a mathematical fractal when your initial guess happens to lie near the boundary between two or more of the equation's roots. The computer becomes caught in its iteration and bucks wildly trying to reach all the roots at the same time, revealing places where Newton's method has broken down into randomness. The pattern produced by this chaotic oscillation is a swarm of spiral shapes, which at different sizes and scales are reflections of each other—revealing that in the space between the roots lurks a fractal, a mathematical strange attractor.

FRACTALS, FRACTALS EVERYWHERE

If such a rich, complex, even creative world can be generated by iterating simple mathematical equations (which are in essence symbolic statements of human logic), could iteration be a key to the creative potential in nature, which has far more interesting things to iterate?

Mandelbrot says, "Fractal shapes of great complexity can be obtained merely by repeating a simple geometric transformation, and small changes in parameters of that transformation provoke global changes. This suggests that a small amount of genetic information can give rise to complex shapes and that small genetic changes can lead to a substantial change in shape." He adds, "The purpose of science has always been to reduce the complexity of the world to simple rules." Is Mandelbrot espousing reductionism?

If so, it's a new brand of reductionism where the simple and complex are interwoven. In that sense it is utterly unlike the old reductionism, which sees complexity as built up out of simple forms, as an intricate building is made out of a few simple shapes of bricks. Here the simple iteration in effect liberates the complexity hidden within it, giving access to creative potential. The equation isn't the plot of a shape as it is in Euclid. Rather, the equation provides the starting point for *evolving* a shape that emerges out of the equation's feedback. Is fractal geometry a better mirror than Euclidean geometry for the order and creativity of nature?

Many of the mathematical fractals that can be generated out of a single, repeated iteration have a wealth of detail, yet they're far too orderly to correspond to natural forms and qualify for Mandelbrot's claim that genuine creativity can lie in iteration and fractals. However, when a random variation in the iterations is allowed so that details vary from scale to scale, it's possible to mimic the actual forms and structures of nature much more closely. This suggests that natural growth is produced through a combination of iteration and chance. But that's only a small part of the story. In the past few years an immense amount has been learned about fractal geometry—and fractals have begun to reveal a great deal about the hidden nature of chaos and order in the natural universe.

The fractal dimension of a Koch island lies between 1 and 2 and corresponds to a ragged curve that shares some of the properties of a two-dimensional surface. But there is also a rich variety of fractals whose dimensions lie between that of a point (0) and a line (1).

For example, fractal structures also appear in the intermittent noise met in period doubling. Here a nonlinear amplifier gives brief barrages of static in the midst of good reception. However, when the static is examined in finer detail, it's found to have layered inside it periods of silence. Then, at even smaller scales, the remaining noise still contains gaps of silence. Intermittency clearly has a fractal structure with the noise/silence detail repeated at finer and finer scales.

The nineteenth-century mathematician Georg Cantor first described this sort of intermittent structure. Cantor, who discovered how to count beyond infinity and create transfinite numbers, became fascinated by the infinite number of points that lie on a line. Suppose, he said, you remove the middle third of a line, then remove the middle thirds of the remaining two lines and go on removing middle thirds ad infinitum. The result is a "discontinuum," a dust of points. Mandelbrot has likened this "Cantor dust" to the gaps that form when milk curdles.

Cantor dust has a fractal dimension of 0.6309, lying midway between a line and a point. The Cantor set brings to mind the motion paradoxes of the Greek mathematician Zeno. Cantor dust is at one and the same time

FIGURE 0.22. **The Cantor set on the way to Cantor dust.**

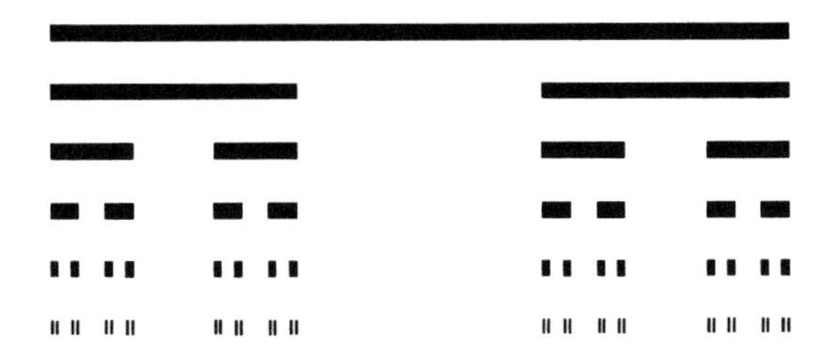

infinitely divisible yet discontinuous. Mandelbrot has claimed—and some physicists agree—that Cantor sets may help to describe the nature of the night sky where the clustering of stars, with corresponding gaps, occurs on many scales, right up to the superclusters (clusters of clusters of galaxies). Current analysis of the structure of the universe suggests a fractal dimension somewhere between one and two. Mitchell Feigenbaum has suggested that by examining the fractal dimension of the present state of the universe, scientists may someday be able to deduce what the universe was like at the beginning.

Turbulence has been described by the torus that breaks apart into a series of fine points. This torus turns out to be a Cantor dust with fractional dimension. Mandelbrot highlights the fractal nature of turbulence by pointing out that in the real world it comes in gusts; it is intermittent. On a stormy night the wind will suddenly drop, then lift again, circling and blowing leaves, then allowing them to fall. This intermittency of turbulence may recur on smaller and smaller scales. Scientists have noticed, for instance, that when a wind tunnel is switched on, the first turbulence that is created isn't stable. At first it fluctuates, and only settles down after the giant fans have been rotating for some time. Does this suggest that the fractal structure of spatial turbulence has another fractal structure that varies in time? (More about fractal time later.)

The weather patterns that Lorenz discovered to be chaotic are also now believed to be fractal. Shaun Lovejoy of McGill University thinks that the atmosphere has a multiplicity of different fractional dimensions. Lovejoy says the problem with weather prediction is not just that the tiniest lack of information about current conditions (the flapping butterfly) accumulates to overwhelm the calculation. He says the meteorologist's weather-gathering network itself has a lower fractional dimension (1.75) than the clouds and winds and other forces it studies. So a deeper problem is that the meteorologist can never get the right kind of data.

Today physicists, economists, biologists, geographers, astronomers, electronics engineers, and anatomists are discovering how an immense number of diverse shapes can be characterized by their fractal dimensions. Everything from the winding of rivers to the convolutions of human brains, from the structure of galaxies to the patterns of metal fractures yields to the fractal measure.

The brains of small mammals are relatively smooth while those of humans are highly convoluted. There appears to be a characteristic fractal dimension of 2.79 to 2.73 for the human brain. Fractal structures are also found in the membranes of liver cells. The nasal bones of the deer and arctic fox maximize their sensitivity to smell by packing the largest possible surface area into a small volume. The result is a fractal structure with constant fractional dimension.

The branching of a living tree is clearly fractal; branches have smaller branches with details being repeated down to the dimension of tiny twigs. One approach to mimicking trees on the computer involves neglecting the thickness of the branches and seeing what happens if the same branching angle is preserved at smaller and smaller scales. The method allows modelers to reproduce a variety of "trees,"

FIGURE 0.23. British scientist Michael Batty's generation of a fractal tree by computer. Each branch splits into two to create a fractal canopy. By the thirteenth iteration (*below*) the tree begins to look more realistic.

Fractal trees illustrate the point that fractal geometry is a measure of change. Each branching of the tree, each bend in the coastline, is a decision point. The decision points can be examined in finer and finer scale, each scale having further decision points.

The fractal structures of real trees are also determined by physical constraints—for example the requirement that each branch must be strong enough to support the weight of the wood it carries, the need to store food in the branches, to drain rainwater, and to avoid excessive wind resistance. Where a number of different constraints are present, a single fractal is insufficient to describe the complexity of the final form. A tree created through iteration of a single equation may look complex but it is clearly mechanical. Fractals become more "organic" when, at each step, there is choice between several alternative forms of iteration, or when a particular fractal iteration persists for several length scales and then suddenly changes.

ranging from cauliflower and broccoli to more familiar trees in which the fine twig structure seems to fill all the available space without actually overlapping. Fractal modelers can produce different species of trees by changing the fractal number.

But real trees have thick trunks, so not only must the lengths of the branches be scaled, their thicknesses must be scaled as well. Leonardo da Vinci noticed that branches grow progressively thinner in such a way that the total thickness (putting all the branches together) above any point is equal to the thickness of the branch below.

FIGURE 0.24. Frost crystals are examples of the fractal patterns that surround us in nature.

Take, for example, the human circulatory system, that amazing piece of engineering consisting of a supply system (arteries carrying oxygen-rich blood) and an exhaust system (veins carrying away waste products). These two systems of branching pipes come from a central pumping area (the heart) and must be arranged in such a way that no part of the body, no organ or piece of tissue is far from both systems. These severe constraints dictate a fractal branching structure for the veins and arteries. However, blood itself is a very expensive commodity in terms of the body's resources; consequently blood has a volume of only 3 percent of the body. The problem is how to get the circulatory system infinitely close to each body part and keep the total blood volume low. Nature's solution is a more rapid branching than simple scaling would suggest. The blood supply bifurcates between eight and thirty times before reaching each particular location in the body and has an overall fractal dimension of 3.

The lung is a particularly illuminating fractal structure and tells us something about the meaning of scaling. What is scale? The ancient Greeks derived history's most famous scale, the golden mean, or golden section. Draw a line and divide it so that the two segments *b* and *a* are in the same ratio to each other as the longer segment is to the whole line. The proportion of *a*/*b* is equal to the irrational number 1.618 . . .

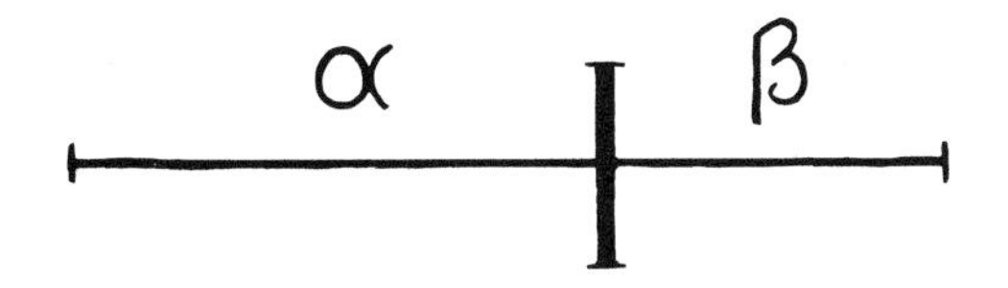

This proportion can also be found in a series of numbers beginning with 1, where each number is the sum of the two preceding it: 1,1,2,3,5,8,13,21 . . . The ratio of each number to its predecessor approximates the golden mean. This series, called the Fibonacci numbers, was named after the thirteenth-century Italian mathematician Filius Bonacci, who made it famous.

Studies have shown that the ratio of lengths in the first seven generations of the human lung's bronchial tubes follow the Fibonacci scale. The diameters of the tubes are classical, that is Fibonacci, up to ten generations. But after these initial generations, the scales change markedly.

Bruce West and Ary Goldberger have demonstrated that the lung incorporates a variety of fractal scales. This shifting of scales allows the lung greater efficiency. For example, after the twentieth iteration the branching takes place at a smaller scale of length but with the same windpipe diameter as the previous iteration. West and Goldberger say, "The final product, which we have dubbed 'Fractional/Fibonacci lung tree,' provides a remarkable balance between physiological order and chaos."

Fractal self-similarity pervades the bodies of organisms, but it is not the blatant homunculus self-similarity that was imagined by earlier science. The body is a weave of self-similar systems like the lungs, the vascular system, the nervous system.*

The bronchial tree is not only a fractal product, it is also a "fossil" of the developmental process that produced it. The time in which the lung grew must also have contained different scales. Is time self-similar and yet random, chaotic? Does it crackle and change scales in its iteration like the bronchial tree?

Our most intimate time clock, the beat of the heart, follows a fractal rhythm. Each beat is essentially the same as the last, but never quite the same. Disruption in the normal fractal scaling of the heart's time can cause pa-

* It's necessary to look at the whole body to see this subtle self-similarity. The immune system and the brain are two very different systems; each has its own fractal dimensions. Yet if Nobel laureate Gerald Edelman is correct, the way the brain sorts out which of its cells will respond to input is a mirror of the way the immune system sorts out which variation of immune cells will respond to a particular disease. For a discussion of Edelman's theory see *Chapter 2*, pages 173–74.

thology in two directions. If the heartbeat and respiration become highly periodic (regular), they can lead to congestive heart failure. On the other hand, a rhythm that is too aperiodic causes the defibrillation of a heart attack. Thus the normal "time" of the heart oscillates in the borderland between order and chaos.

Similarly, in healthy people the counts of a type of white blood cell called neutrophils fluctuate fractally. But with chronic leukemia neutrophils rise and fall in predictable cycles. West and Goldberger conclude that systems iterating in fractal rhythms are normal for the body and that "a loss of physiological variability in a variety of systems appears to be characteristic of the aging process." To be healthy is to be composed of simmering cycles of fractal time.

In the past we have thought of time as an inflexible yardstick to hold against change. But is time itself evolving and shifting like a turbulent stream? Is time a strange attractor? Perhaps this is why psychological time seems to stretch or squeeze like rubber, some moments seeming to fly by, others dragging out. Strange attractors have a self-similarity. Could this be why history seems both to repeat itself and never repeat itself?

In the real world the intermixing of fractals unfolding at different scales gives richness to natural forms and to the time they evolve in. Similarly, the fractal measure has been made richer and more useful by the introduction of the concept of the "random fractal." Here a variety of generators are used which can be chosen at random at each scale. Random fractals not only have an intricacy of detail but also a freshness and unpredictability characteristic of real systems. By combining an iterative scaling with a random element of choice, coastlines, mountains, and planets can be generated that are realistic enough (though completely imaginary) to be used in movies, videos, and advertising.

FIGURE 0.25. Michael Batty generates this "planetrise" by displacing the midpoints of triangles randomly (moving the midpoint right or left on a new triangle) as the iteration process proceeds.

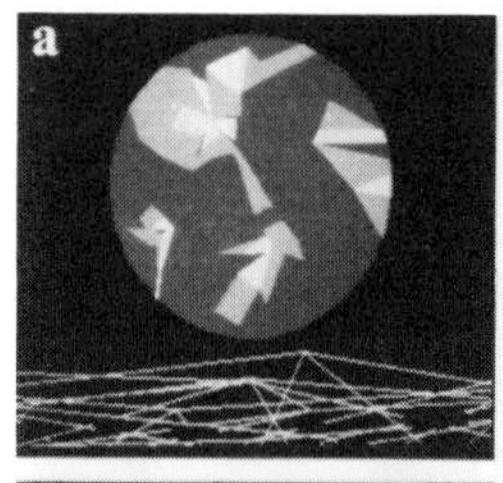

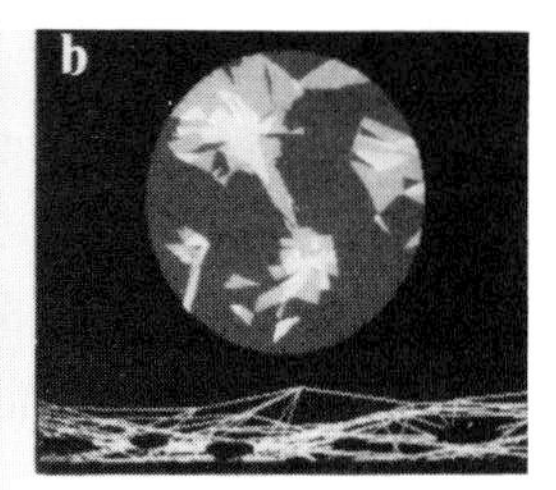

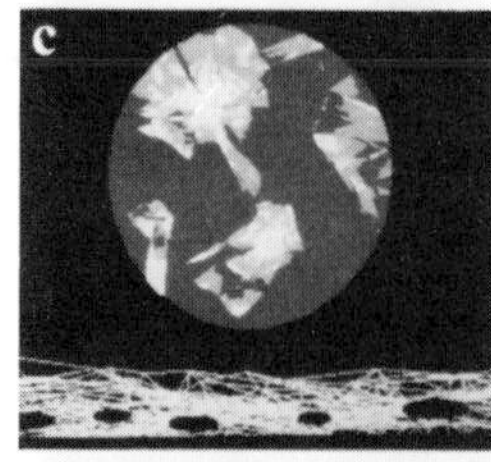

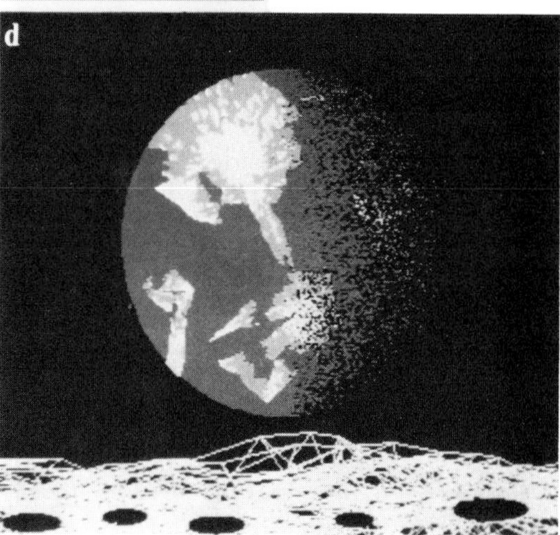

Random fractals appear closely related to a variety of materials such as polymers and solid surfaces. In fact, with the exception of single crystals, most of the materials around us are to some extent disordered. Until the concept of random fractals was invented, it was extremely difficult to describe the appearance and properties of these regular-irregular solids. Now not only their physical form but the processes by which they grow have been modeled by fractal geometry. In fact, random fractals have yielded the ability to model a huge variety of systems. The wake of a supersonic jet plane, the swirling Gulf Stream with its endless subdividing and reuniting branch streams, percolating oil through sand, neural networks, and the spread of a forest fire have all appeared in realistic forms on computer screens as a result of this mathematics.

A very curious wrinkle in the effort to mimic nature through the mathematics of random fractals combines fractals and topology. A team of scientists at the Georgia Institute of Technology led by mathematician Michael F. Barnsley has discovered a wonderfully clever trick for reproducing even very complicated forms realistically by a process called "affine transformations."

Imagine outlining a full-size leaf on a sheet of stretched rubber and then shrinking and skewing the picture into a smaller distorted version of the original. The affine idea is to find several of these smaller leaf transformations that can be overlapped into a collage that has the shape of the original full-size leaf.

Let's say there are four smaller distorted versions of the leaf that can be shingled to make the shape of the full-size original—four affine transformations. Each transformation is a mathematical formula indicating the degree and extent of distortion of the coordinates (outline) of the original leaf.

Starting at some point on the computer screen, Barnsley recreates the original leaf using only his affine transformations and iteration. He first applies one of the transformations to a point. The transformation specifies a point at another spot. He applies another transformation to that spot, and so on. The random iteration of the set of four affine transformations generates a fractal attractor that looks like the original leaf.

It is chance that determines the moment-by-moment application of the affine rules, but the limit of the process has been set by the four transformations describing the original leaf. Therefore the original leaf will always appear when the iteration is carried out.

Implications that affine transformations might have for morphogenesis (the way form develops in living organisms) in the real world are yet to be explored. But in immediate practical terms, scientists are hoping these transformations will allow them to derive efficient ways to store complex data in digital memory, transmit photographs over phone lines, and simulate natural scenery by computer.

So in everything from new forms of image animation to the abstract mathematics of strange attractors to the geometry of a head of broccoli, Benoit Mandelbrot's fractals are infiltrating our perception of the world.

Their appeal is immediate. Richard Voss, Mandelbrot's colleague at IBM and himself a prolific creator of fractal landscapes, says, "I get many, many letters from people who say they couldn't care less how these things are drawn, but that the shapes are beautiful or scary or attractive or repulsive. Mathematics has gotten much closer to these people's daily experience and emotions."

David Ruelle makes a similar point in his seminal paper, "Strange Attractors."

"I have not (yet) spoken of the esthetic appeal of strange attractors. These systems of curves, these clouds of points, suggest sometimes fireworks or galaxies, sometimes strange and disquieting vegetal proliferations. A realm lies here to be explored and harmonies to be discovered."

But is this really the way nature is? Peitgen and Richter point out that fractal pictures "represent processes which are, of course, simplified idealizations of reality. They exaggerate certain aspects to make them clearer. For example, no real structure can be magnified repeatedly an infinite number of times and still look the same. . . ." In nature itself after only a few iterations a new order takes over.

But fractal geometry isn't meant to be an exact representation of complexity. In fact, that's the point.

The ancient Greek philosopher Anaximenes has been called the father of science because he was the first to propose that the different qualities in things are caused by different quantities of their elements. Using quantitative differences to account for qualitative ones has been a hallmark of science ever since. Fractal geometry, like Thom's catastrophe theory and the other measures of change, involves a massive shift in that ancient tradition.

The scientists of change have learned that the evolution of complex systems can't be followed in causal detail because such systems are holistic: Everything affects everything else. To understand them it's necessary to see into their complexity. Fractal geometry abundantly provides that vision: a picture of the *qualities* of change.

It may initially feel "unnatural" to see in this way because our perceptions of the world are still very much influenced by the aesthetics of the Greek philosophers and the notions of Platonic ideals and Euclidean forms. We are used to picking out such shapes as parallel lines, circles, triangles, squares, and rectangles in nature or in art. We accept it as obvious that music and art should be based on basic symmetries and relationships. However, as the inner nature of chaos and the complex and subtle orders of living systems—flowing rivers, rotating galaxies, light and sound, growth and decay—reveal themselves to our scientific perception, we will begin to realize how static and limited the Platonic and Euclidean ideas are. Regular, simple orders are in fact exceptions in nature rather than the rule. Nature's true archetypes may well lie closer to Ruelle's strange attractors and Mandelbrot's fractals than to Platonic solids.

Strange attractors and fractals evoke a deep recognition, something akin to the haunting recognition afforded by the convoluted and interwoven figures of Bronze Age Celtic art, the complex designs of a Shang ritual vessel, visual motifs from the West Coast American Indians, myths of mazes and labyrinths, the iterative language games of children or the chant patterns of so-called "primitive" peoples. The regular harmonies of classical Western art become almost an aberration set beside these forms. Yet as we look at the greatest art we realize that even in classical forms there is always a dynamism of chaos within the serenity of order. All great art explores this tension between order and chaos, between growth and stasis. In confronting the orders of chaos, of growth and stability, it appears we are now coming face to face with something that is buried at the foundations of human existence.

The depth of that encounter is suggested by the pioneering work of psychiatrist Montague Ullman and others, which indicates that even the structure of our dreams may be fractal. Researchers believe that the dream "story" contains repetitions of the dreamer's central concerns. Reflections of these concerns can be found in both the overall "story" and in its finer and finer detail.

Perhaps some of the appeal of the fractal is that in each of its "parts" it's an image of the whole, an image in the looking glass.

FIGURE 0.26. Self-similar pattern on the Desborough mirror made by Celts, probably sometime in the first century A.D.

A few years ago, looking-glass physicist David Bohm proposed another scientific image to convey a new holistic view of nature: the hologram.

A hologram is made by shining laser light (light of a single wavelength) through a half-silvered mirror. Half the laser beam is directed onto a photographic plate. The other half is bounced off an object and *then* onto the plate. The two halves of the light meet at the plate and interfere with each other. The interference pattern is recorded on the plate and looks like a fine-grained picture of the wave pattern created by pebbles thrown into a pond.

When a laser beam is later directed through the plate, an image of the object photographed unfolds from the wave pattern and projects three dimensionally in space. A viewer can walk around this chimerical object and see it from different perspectives just as one would see a real object. The whole object has been recorded in the interference pattern. Cutting a piece from the hologram and sending the laser beam through the fragment also produces an image of the whole object, although this image may not be quite as sharp. This holistic effect is analogous to the self-similarity of a fractal, repeating the shape of the whole at different scales.

Bohm uses the hologram to illustrate his contention that light and energy and matter all over the universe are composed of moving interference patterns which literally bear the mark of all the other waves of light and energy and matter they've been in contact with, directly or indirectly. In other words, each part or instance of energy and matter encodes an image of the whole.

For Bohm, holograms describe the deep construction of matter and movement of energy. Mandelbrot's fractals describe the shapes that matter takes and the orderly and chaotic processes that transform those shapes. Both seem to suggest that each part or phenomenon in the physical world represents a microcosm of the whole.

Fractals are a descriptive system and a new methodology for an investigation that has only just begun. They may also be, like the hologram, a new image of wholeness. Over the next decade fractals will undoubtedly reveal more and more about the chaos hidden within regularity and about the ways in which stability and order can be born out of underlying turbulence and chance. And they will reveal more about the movements of wholeness.

In his painting "The Great Wave," the eighteenth-century Japanese painter Katsushika Hokusai beautifully captured all these aspects of the fractal world we're about to enter.

FIGURE 0.27

The Yellow Emperor went wandering north of the Red Water, ascended the slopes of K'un-lun, and gazed south. When he got home, he discovered he had lost his Dark Pearl. He sent Knowledge to look for it, but Knowledge couldn't find it. He sent the keen-eyed Li Chu to look for it, but Li Chu couldn't find it. He sent Wrangling Debate to look for it, but Wrangling Debate couldn't find it. At last he tried employing Shapeless, and Shapeless found it.
The Yellow Emperor said, "How odd!—in the end it was Shapeless who was able to find it!"

CHUANG TZU

0 CHAPTER

We've passed through the mirror portal. Here, on the other side, everything looks different.

In the landscape on the first side of the mirror we saw how simple, apparently stable systems can iterate into chaos. In the terrain on this side we'll see how chaos gives birth to stable order. Here we'll behold strange sights such as magical waves that can travel for thousands of miles without changing shape; we'll find out about feedback and cooperation as a new concept of evolution; and we'll glimpse the secret order of art. On this side of the mirror we'll see how iteration, bifurcation, critical values, fractals, and nonlinearity apply not just to disintegrating systems but to emerging ones, from vortexes to stars to human thought.

Even our view of the Yellow Emperor will change. On the other side of the mirror he seemed to be the keeper of reductionism, but here he seems to have known about the more holistic view all along. Perhaps it's because he's a Taoist that he doesn't mind such contradictions.

A. A *violent order is disorder: and*
B: A *great disorder is an order.*
These two things are one.

WALLACE STEVENS
"CONNOISSEUR OF CHAOS"

Chaos TO ORDER

Chapter 4

The air finds its way in everywhere, water passes through everything.

"THE YELLOW EMPEROR"
LIEH-TZU

JOHN RUSSELL'S OBSESSION

Throw a stone into the center of a lake and the disturbance soon spreads out and dissipates. Try to mold the water in your bathtub into the form of a little hill and it flows away as fast you you can gather it together. It's in the nature of waves to break up.

That's why the experience of Scottish engineer John Scott Russell one day in August 1834 was so remarkable. Russell was riding his horse along the Union Canal near Edinburgh when,

> I was observing the motion of a boat which was rapidly drawn along a narrow channel by a pair of horses when the boat suddenly stopped—not so the mass of water in the channel which it had put in motion; it accumulated round the prow of the vessel in a state of violent agitation, then suddenly leaving it behind, rolled forward with great velocity, assuming the form of a large solitary elevation, a rounded, smooth and well-defined heap of water, which continued its course along the channel apparently without change of form or diminution of speed. I followed it on horseback, and overtook it still rolling on at a rate of some eight or nine miles an hour, preserving its original figure some thirty feet long and a foot to a foot and a half in height. Its height gradually diminished, and after a chase of one or two miles I lost it in the windings of the channel.

Russell was a trained engineer and ship designer. He knew how unusual it was to see a wave continuing on its path at a constant speed and shape, never falling into a flurry of foam, never dividing into many smaller wavelets, never losing its energy, but rolling on until he could no longer follow it.

That unnatural wave, which today is known as a "soliton," or solitary wave, preoccupied, obsessed, and mystified Russell for the rest of

Figure 4.1. Adding together sine waves to portray a more complicated wave.

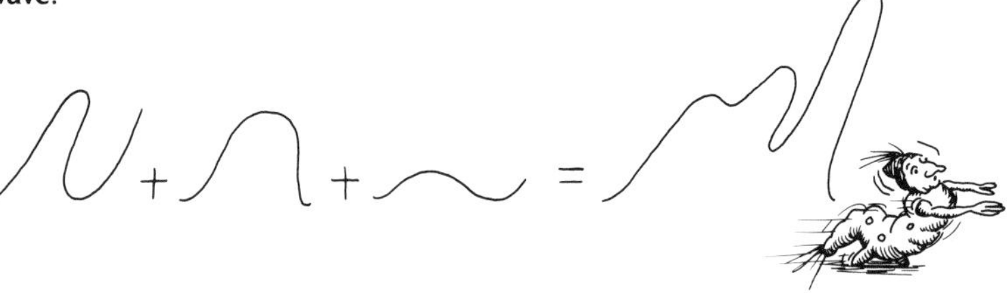

his life. It was to become the basis for his revolutionary design of ships' hulls. In our time, it became one of the most important new concepts to sweep across the sciences.

To understand what is so remarkable about the soliton wave, we need to examine in a little detail what happens to an ordinary wave in a very deep canal.

Physicists have devised a technique that allows them to picture any complicated shape such as a wave as made out of a combination of sine waves. A sine wave is the simplest form a wave or oscillation can take. Each sine wave is characterized by its frequency or its number of vibrations each second. When several simple sine waves are added together, they produce a more complex shape. An electronic music synthesizer works on this principle. The synthesizer can reproduce the sound of any musical instrument by adding together the outputs of several pure sine wave oscillations, each of a different frequency.

The lump of water making up a wave on the surface of a canal can be described as made up of a number of sine waves, each of a different frequency. In water, waves of different frequencies travel at different speeds. Because there is nothing to hold these different frequencies together, the lump of the complex wave changes shape; its crest begins to peak and overtake the main body. The breaking up of waves into smaller disturbances and finally chaos is known as dispersion. Waves suffer dispersion because in a linear world individual sine waves are independent of each other. But clearly dispersion did not occur to the wave seen by John Russell. Why?

Scientists now know that the wave Russell observed owed its stability to nonlinear interactions binding the individual sine waves together. These nonlinearities took place near the bottom of the canal and caused individual sine waves to feed back into each other, creating the reverse of turbulence. Instead of smoothly oscillating water becoming increasingly fragmented, at a critical value the sine waves coupled. As one sine wave tried to speed up and escape from the soliton, its interaction with the others held it back.

Think of a marathon race in which thousands of runners are all bunched together at the start. As soon as the race starts, runners begin to separate and after a short while the pack breaks apart. This is just what happens to an ordinary wave. A soliton wave, however, is like the best runners in the race. Mile after mile they remain coupled by feedback. As soon as one tries to pull away, the others compensate and the group stays intact.

A soliton is born on the edge. If too much energy is involved in the initial interaction, the wave breaks up into turbulence. If too little energy, the wave dissipates. On this side of the mirror, nonlinear interactions at critical values don't produce chaos, they produce spontaneous self-organizing forms.

Russell didn't know why his solitary wave formed, but he was soon at work building an experimental wave tank in his garden and working with barges on the canal. He quickly discovered how to generate what he called "waves of translation" at will and noticed that their speed was always related to their height. This meant that it was possible for a tall, thin wave to chase after and catch up with a short fat one. He also discovered that the existence of these waves was connected to the depth of the canal. If the Union Canal had been much deeper, he might never have seen his soliton.

With considerable foresight Russell realized that the significance of the wave of translation must extend far beyond the Union Canal. He was able to use principles of the wave to show that the sound of a distant cannon is always heard *before* the order to fire because the cannon sound moves as a soliton or solitary wave and so travels faster. Using the soliton principle he was able to correctly calculate the depth of the atmosphere, and he even tried to use it to determine the size of the universe. At the time of his death in 1882 Russell was working on a book, *The Wave of Translation*, which was posthumously brought out by his son.

Russell's contemporaries found little merit to all this work. They believed that his obsession with his wave of translation had led him into what one critic called "many extraordinary and groundless speculations." Textbooks on wave motion published in the last century made only passing references to Russell's oddity.

Ten years after Russell's death, however, the Dutch mathematicians D. J. Kortweg and C. de Vries wrote down the nonlinear equation, called the KdV equation, which contains Russell's wave as its solution. But this, too, had little impact. While acknowledged as a useful piece of mathematics, it wasn't thought to have much importance to the rest of physics.

The KdV equation confirmed Russell's observations about what happens when two soliton waves collide. This is backed by modern water tank observations and computer modeling. A high, thin, humpbacked soliton catches up with its fatter cousin and the two waves meet and coalesce for a time. What happens next is quite astonishing. The combined

Figure 4.2. A short soliton passing through a tall one.

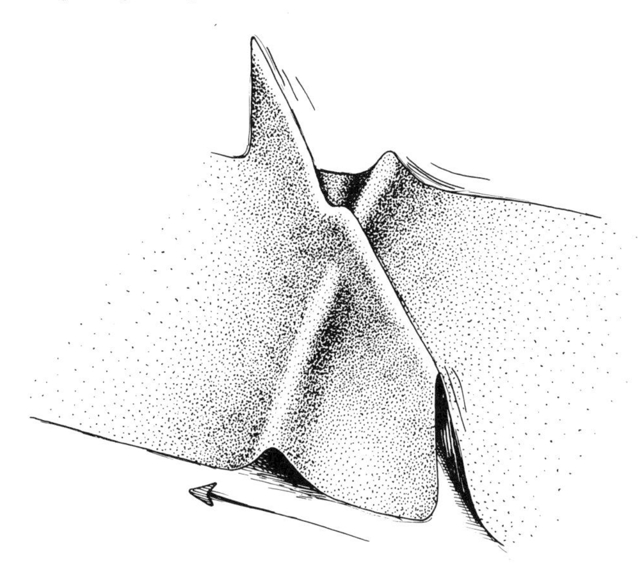

soliton separates so that the faster, higher wave travels on at its original rate of motion, leaving the short fat wave behind. Speeded up, the event looks as if the faster wave simply passed through the slower one like a Hollywood special effect.

Where the two soliton waves cross, there is no separation of one wave from the other, yet the two waves emerge intact. Could this indicate that there is a kind of memory in the nonlinear couplings where the waves remember their former order? We've seen a nonlinear memory before in intermittency.

The KdV equation also describes a relative of Russell's soliton, the tidal bore. This phenomenon is also called a "whelp," "stubble," or "mascaret." In the River Severn in Britain, exceptionally high tides force a mass of water through the funnellike river mouth and then up the gently sloping estuary. When the difference between high and low tides reaches around 20 feet, a very large mass of water is ejected into the river where the sloping bed acts to focus the surging water into a soliton. One result of this tidal bore is that the actual flow of the river is reversed and water begins to flow upstream.

In the Amazon River bores 25 feet high have been known to travel for over 500 miles. In Nova Scotia's Bay of Fundy bores reach 30 feet in height. Ranging from a few inches to walls of water tens of feet high, tidal bores are found all over the earth.

MORE WAVES AND A RED SPOT

In the 1970s, after solitons were dug out of the scientific closet and became the rage, researchers explored other water nonlinearities. Henry Yuen and Bruce Lake from the TRW Defense and Space Systems Group realized that back in 1890 the famous mathematician and physicist Sir George Stokes had made a valuable contribution to the subject. Stokes' theoretical investigations indicated that it was an oversimplification to apply the principle of lin-

Figure 4.3. **A tidal bore forming.**

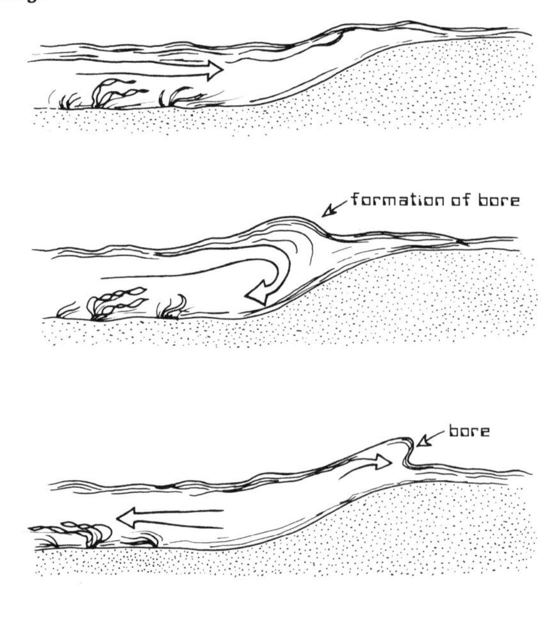

ear dispersion to very deep water because in the ocean nonlinearity has to be taken into account. For example, the effects of gravity must be considered at various depths. By using Stokes' nonlinear terms, Yuen and Lake derived a cousin of the KdV equation; it shows that while solitons in deep water don't exhibit the simple relationship between height and speed observed in Russell's soliton, they nonetheless can move for great distances without any change of shape, and can survive collisions. Photographs of these solitons have been taken by satellite. (See *Figure 4.8.* Remember which side of the mirror you're on; the figure you're looking for is on page 133.)

The world's most violent water soliton is without a doubt the tsunami, or seismic wave, misnamed "tidal wave" (it has nothing to do with the tides). Although the tsunami occurs in the ocean, it is treated mathematically the same as a wave in shallow water, the wave in a canal. This is because the enormous wavelength (that is, the distance between the crests) of the tsunami may be hundreds of meters in length, a dimension which far exceeds the depth of the ocean.

Tsunamis are formed when a strong seismic shock occurs on the ocean floor. The wave, only a few inches or feet high, can travel intact across the ocean for many thousands of miles. Because of its very long wavelength, it may take as much as an hour for a complete wave to pass a particular point. An ocean liner that goes through a tsunami will only experience a gentle rise over a period of tens of minutes—undetectable except by the most delicate instruments.

The human problem begins when the tsunami reaches the continental shelf. In shallower waters, nonlinear effects at the seabed act to shorten the wavelength of the wave and increase its height. The result is awesome. From a soliton a few inches or feet high, the tsunami becomes a 100-foot mountain of water crashing into coasts and harbors. The tsunami that killed thousands in Lisbon in 1775 caused many writers in the Age of Enlightenment to question the existence of a benevolent God. In 1702 a tsunami in Japan drowned over 100,000 people, and in the seismic soliton created by the volcanic explosion of Krakatoa Island in 1882 thousands died.

If solitons are produced in water, why not in air? Could there be stable pulses in the atmosphere that propagate undisturbed over great distances?

What may be the first recorded atmospheric soliton was the mass of cold air that moved across Kansas on June 19, 1951. A sudden change in air pressure proceeded along a temperature inversion at a height of over a mile. Records show that the soliton front was more than a hundred miles long and traveled at about 12 miles per hour for several hundred miles. A pulse of such stability and constancy must have been the result of nonlinearities that coupled together the disturbances of the atmosphere and prevented them from dissipating.

In recent years, meteorologists have made a close study of atmospheric solitons and learned that they occur in two forms. One is called an E-soliton or "wave of elevation," analogous to Russell's water wave. The other is the D-soliton or "wave of depression," a sort of *anti*soliton.

These soliton waves have been observed not only in our own atmosphere but also in the atmospheres of other planets. Near the Tharis Ridge on Mars, the properties of the atmosphere change slightly in the morning hours of late spring and early summer. The result is a borelike disturbance that sweeps along the ridge.

Perhaps the most well known soliton of any sort is on Jupiter.

In 1664 the English scientist Robert Hooke observed a reddish spot on the surface of the

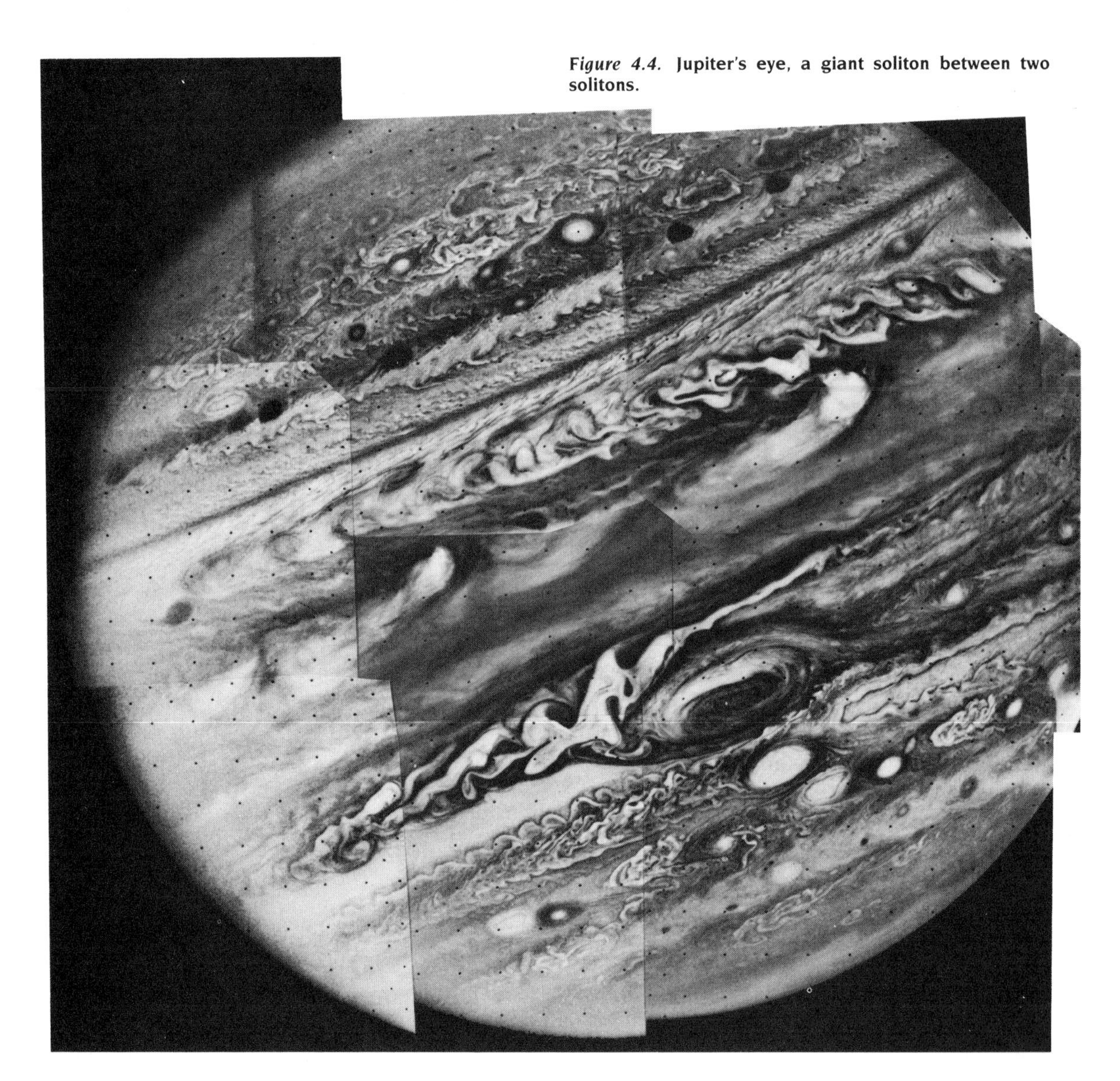

Figure 4.4. Jupiter's eye, a giant soliton between two solitons.

giant planet. Further sightings were made over the next fifty years, but between 1713 and 1831 no other observations were set down. Throughout the second half of the nineteenth century, however, this feature of the planet's atmosphere became increasingly prominent.

The spot is located in the southern hemisphere just below the planet's equator and is so big that the entire earth could be comfortably dropped into it. Ideas that the giant red spot was a mountain peak or plateau were discounted when scientists realized that the surface of Jupiter isn't solid; it's made of compressed, liquefied gases. Another theory, that the spot was a raft of ice, was also ruled out. The Great Red Spot, as it came to be known, must be atmospheric because, while it maintains a constant latitude, its longitude changes as the spot moves around the planet. A large disturbance of such longevity posed a considerable puzzle to planetary scientists. Other spots have also been discovered, several on Jupiter and one on Saturn that is a quarter of the size of Jupiter's.

How could an atmospheric eddy remain stable for centuries? The prevailing winds above and below the Red Spot travel in opposing directions at some 100 meters per second, but the spot moves only a few meters per second. The spot is held between two high-velocity air streams like a ball bearing rolled between two hands.

In 1976 two scientists from the University of California suggested that the Red Spot is an enormous E-soliton, a nonlinear wave of elevation that is trapped between two D-solitons.

According to this model, the Great Red Spot is not very deep; it floats on top of the Jovian atmosphere. A boost for the soliton nature of the Red Spot came from observations of the South Tropical Depression, another prominent feature of Jupiter's atmosphere, which has been known for several decades and appears to be a D-soliton. To the surprise of astronomers, in the 1950s this disturbance approached the red spot, appeared to enter into it and vanish, only to slide out intact and unchanged on the other side. In the linear world such behavior would be totally unexpected, but it is everyday nonlinear magic.

With more sophisticated computing power scientists have been able to reproduce this magic in both models and experiments. In 1988, Philip S. Marcus of the Department of Mechanical Engineering at the University of California tested out one theory of giant soliton vortices on Jupiter, producing an animated computer film that showed how small vortices form spontaneously and, given the right conditions of shear winds in the Jovian atmosphere, are swept up into a larger and more stable spot. In the words of Andrew P. Ingersoll, one of the creators of this theory: The impression is of large scale order spontaneously emerging out of small-scale chaos.

Following on Marcus's graphic demonstration, three scientists at the University of Texas, Austin—Joel Sommeria, Steven D. Meyers, and Harry L. Swinney—tried to see if they could actually produce a "red spot" in the laboratory. In order to create the sort of shear that takes place on Jupiter, the three used a rapidly rotating cylindrical tank into which fluid could be pumped from an inner ring and withdrawn from an outer ring. When the right pumping rate and speed of rotation of the cylinder was attained, some of the fluid began to rotate in the opposite direction and produce a region of shear where vortices appeared, began to merge and formed a much larger and stable "red spot." On the giant planet itself, while friction will constantly drain away energy, convection currents constantly carry new fluid in and out of the Red Spot. In addition the Spot acts to absorb any smaller vortices that happen to form in its vicinity.

SOLID SOLITONS

Though the inside of solid metal seems an unlikely place to find solitons, a 1955 study of electron movement in vibrating metal lattices brought the whole soliton topic to the attention of scientists.

Interest in lattice solitons came from a rather academic little problem, the question of the "equipartition of energy."

One of the cornerstones of physics is the field of statistical mechanics, which deals at the molecular and atomic level with the relationships of energy to change. Statistical mechanics is the key to thermodynamics and describes almost every change in nature, from those occurring in a living cell to those in your car engine. A central assumption of statistical mechanics is the principle of equipartition or democracy of energy.

Equipartition describes what happens when a system is given a little extra energy, a bonus packet of heat, for example. Scientists have always assumed that this energy will rapidly become distributed over the whole system. This makes equipartition a little like a rich man going into a crowd of pickpockets. Sooner or later everyone will have picked everyone else's pocket and the money will be evenly distributed over the crowd. This principle explains why things always move toward equilibrium, why heat at one end of a fire iron begins to distribute itself, and why an initially active system eventually runs down.

Whenever energy is localized or concentrated in a particular part of a system or is associated with a particular activity, then that system has the potential to change itself and do work. But according to the principle of equipartition, this energy will also tend to dissipate. From the energy's point of view there are no privileged places—everywhere is like everywhere else. Since work and activity require a flow of energy from one place to the next, when energy becomes the same everywhere all activity will die down.

The notion that energy will eventually become equally distributed in all systems was proposed in the middle of the last century and was universally accepted. However, because of the difficulties involved in actually calculating the behavior of a large number of molecules in a system, it wasn't possible to follow the fine details of equipartition and see how energy would pass from molecule to molecule. Scientists were forced to assume the principle was true.

With the development of computers, however, scientists could peer into the way energy migrates through a crowd of molecules. In the mid-1950s the noted physicist Enrico Fermi, aided by the mathematicians Stanislav Ulam and J. Pasta, decided they would look at the vibrations in metal using the then state-of-the-art computer, Maniac I.

The internal structure of metal contains a stable pattern, called a lattice, of atoms. When energy, in the form of heat, is given to the metal it causes the atoms to vibrate. But because these atoms are all bound together in the lattice they vibrate in a collective way, producing a single "note." In fact, there are many notes, many different modes of vibration within the lattice, and each of these is associated with a characteristic energy.

According to the principle of equipartition, if all the heat energy were to be given to a certain note—that is, to a particular vibration of the lattice—then pretty soon that energy would spread out and distribute itself to all the other "notes" of the lattice. This was the great assumption of thermodynamics and, since no one could actually get inside a lattice to see what was happening, it had never been observed directly. But with the coming of the computer the lattice could be looked at indirectly, through a mathematical model.

To observe the way energy was shared between all the vibrational notes in the lattice, Fermi, Pasta, and Ulam set up a model containing five notes or modes. The plan was to feed one mode with energy and watch how this energy obeyed the strictures of thermodynamics by distributing itself through the other modes. In order to mathematically represent this sharing of energy it was necessary to add a tiny extra term—a nonlinear term—corresponding to the interaction between modes. If it was not added there was no way that "energy" in the model could pass from one note to another. As it turned out, this tiny additional term dominated the whole system and transformed it from a linear, well-behaved lattice into an arena for solitons.

In the 1950s when the Fermi-Ulam-Pasta calculation was carried out no one was seriously thinking about solitons, so the three scientists were quite confident that once the system had settled down from its initial burst of energy, the energy would soon be parceled out among all the other vibrational modes.

As expected, after a few hundred cycles of the calculation, mode 1 began to fall rapidly in energy and modes 2, 3, 4, and 5 began to gain. And after 2,500 iterations of the equation everything was still going according to plan. Then something wonderlandish occurred. While vibrational mode 1 continued to lose energy, mode 4 began to gain at the expense of all the other modes. By 3,500 cycles mode 4 had peaked and now mode 3 was beginning to gather energy. To the complete surprise of the scientists, energy was not being shared out equally but was bunching itself together in one or another of the modes. By the end of 30,000 cycles, energy was not equipartitioned at all but had returned and gathered itself again into the first mode!

The result was especially shocking because it was found that this concentration of energy doesn't depend on the strength of the nonlinear interaction; even a very weak coupling of feedback will cause the system to bunch.

The computer calculation indicated that the nonlinear lattice had a sort of "memory" not possessed by its linear counterpart. Given sufficient time, the system would return again and again to the state it was in when it first received its burst of energy—a "Poincaré recurrence." Analysis of the Fermi-Pasta-Ulam model shows that the phenomenon involves formation of a soliton—not of water or air but of energy—which moves through the lattice in a coherent wave.

The model is illuminating because it shows that the nonlinear world is holistic; it's a world where everything is interconnected, so there must always be a subtle order present. Even what appears on the surface as disorder contains a high degree of implicit correlation. Sometimes this below-the-surface correlation can be triggered and emerges to shape the system. Soliton behavior is, therefore, a mirror of chaos. On one side of the mirror, the orderly system falls victim to an attracting chaos; on the other, the chaotic system discovers the potentiality in its interactions for an attracting order. On one side, a simple regular system reveals its implicit complexity. On the other, complexity reveals its implicit coherence.

The ocean soliton is a good example of this implicit coherence. Scientists have always assumed that waves far out at sea are totally random in their shapes and distribution. They believed the sea is so disordered that the appearance of any given wave is a matter of pure chance. However, since nonlinear interactions are always present, the highly complex face of the ocean conceals a subtle form of order that can be triggered in a tidal wave. In the words of Yuan and Lake, the surface of the ocean is "highly modulated" so that it actually contains a remembrance of all its earlier structures. The occasional giant waves that occur in the ocean are now thought not to be fortuitous accidents brought about by the chance meeting of various currents. These giant waves can be considered a self-focusing or surfacing of the ocean's memory in the form of a soliton.

Before the results of Fermi, Ulam, and Pasta, these reflections would have seemed absurd and fantastical. In a lecture given at a conference on nonlinearity at the University of Miami in 1977, Yuan and Lake drew the audience's attention to a quotation from Lewis Carroll's *Through the Looking-Glass*.

> "I can't believe *that*!" said Alice.
>
> "Can't you?" the Queen said in pitying tone. "Try again: draw a long breath, and shut your eyes."
>
> Alice laughed. "There's no use trying," she said. "One *can't* believe impossible things."
>
> "I daresay you haven't had much practice," said the Queen.

Soliton scientists, working in the nonlinear universe, seem to be getting more and more practice.

Following the work of Fermi and his colleagues, scientists stepped up their study of the ways in which vibrations can move through the atomic lattice of a solid. They discovered that a sharp blow to the end of a metal rod will produce a soliton of mechanical energy that travels undisturbed to the other end of the rod. Even a burst of heat will propagate in a coherent wave. Put the end of an iron toasting fork in a warm cup of coffee and heat will slowly diffuse up the iron to the handle. But put the fork into the white-hot center of a camp fire and a ballistic pulse of heat will travel up the metal rod in the form of a soliton.

Scientists now appreciate that whenever a dynamic stability survives, solitons may be present. This led to a realization about one type of soliton that people have been looking straight into for a very long time.

How many scientists had stared into the flame of a candle and asked, why does its ethereal form not die out, or flare up into a sudden burst of light? Michael Faraday said, "All of physics and chemistry is in a candle flame." The miracle is that despite the intense combustion that takes place, the flame always maintains its stability at more or less the same shape and intensity. While Russell's soliton represented a delicate balance of nonlinearity in the realm of dispersion, the candle flame stands for the balance of nonlinear reactions in the realm of diffusion.

For the flame to persist, a new source of energy must flow into it as rapidly as heat and light energy flow out. This means that wax melts, is sucked up the wick by capillary action, vaporizes, and enters the heart of the flame; at the same time oxygen must diffuse into the flame at exactly the correct rate. The soliton as a balance between the inward and outward diffusion of energy is one of the great magic tricks of nature.

BIOLOGICAL SOLITONS

Diffusion solitons are also important in biological systems. Until the development of soliton theories, the problem of understanding how packets of energy are transported down very long molecules was called "the crisis in bioenergetics." In the world of *linear* molecules, energy always tends to spread out so that the right concentration would never get to the exact site. Russian scientist A. S. Davidov asked if *non*linear interactions would help to transport energy along the helical coils of a protein molecule.

Davidov proposed that below a certain threshold, energy is carried by the normal vibrations of the helix backbone and tends to dissipate across the whole molecule. But when a threshold is reached, nonlinearity balances the forces of diffusion and allows a bound packet of energy to travel along the helix at just over a thousand meters per second. In this way, energy which arrives at one part of the molecule can be transported to some other site, where it is used in the cell's biological processes.

Solitons are also employed in biological systems for movement of signals along nerves. If you step out of the ocean onto very hot sand, the sensation of pain must travel along a 5- or 6-foot nerve pathway to your brain. Clearly the information that there is pain at the sole of the foot must not only go a long distance, it needs to arrive intact—no use if a pain message starts out in the foot and ends up in the brain as information about a tickling sensation.

Early workers in the field of nerve conduction knew that signals involved some form of electrical activity. So they came up with a model based on the telegraph or telephone exchange in which messages run along wires.

The only trouble with this theory was that while an electrical pulse in a wire travels close to the speed of light, nerve impulses are far slower, moving at around 32 feet per second.

During World War II great strides were made in electronics and the ability to make fast and delicate electrical measurements. Alan Hodgkin did wartime research into radar but in 1945 returned to his Cambridge laboratory. With the help of his student, Andrew Huxley, half-brother of the famous novelist, Hodgkin began to study the electrical changes that occur in the giant nerve axon of the squid. Their researches clarified that nerve transmission is not at all like the messages in a telephone line; instead it involves a localized pulse that travels down the nerve at constant speed and without changing shape. In addition, each pulse is generated only when a certain critical threshold of energy is reached.

This research earned Hodgkin, Huxley, and John Eccles the Nobel prize. It showed that nerve impulses travel as what we now call solitons, at constant speed and without dissipation. The mathematics of the Hodgkin-Huxley theory revealed that after nerves fire at their threshold, they have a dormant period before another soliton can be generated. The propagation and interaction of neural solitons also involves a "memory." The neuron retains a sensitivity to messages it has passed earlier. In this way, a nerve network has a holistic memory of its pattern of messages, a fact which may be of some significance in developing a general theory of the brain's memory. A whole new area of study has now developed to investigate how solitons collide, pass over irregularities in the nerve fiber, and interact at junctions. Some theoreticians have called the nerve soliton the "elementary particle of thought."

SOLITON TUNNELS

Even a magnetic field can have soliton behavior and here solitons reveal another remarkable feature—the ability to "tunnel."

Ordinarily a magnetic field can go quite easily through a piece of metal. This is why it's possible to hang a nail from the pole of a magnet and then use that nail to pick up another nail. But in a superconducting metal, magnetic "transparency" is suddenly switched off. At the critical temperature, the point at which the metal converts into a superconductor (itself a soliton), the magnetic field finds itself suddenly unable to enter.

Figure 4.5

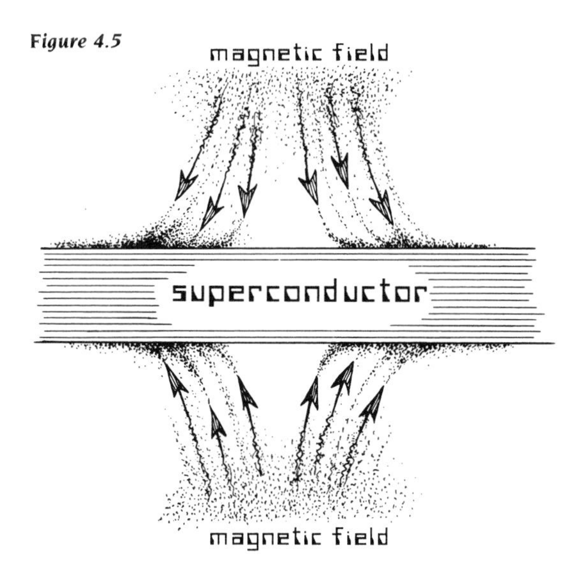

However, if this magnetic field is made stronger and bigger, there comes a point in the field where solitonlike vortices of magnetism are created that penetrate or tunnel right into the superconductor. In effect, it is one soliton passing through another.

Figure 4.6

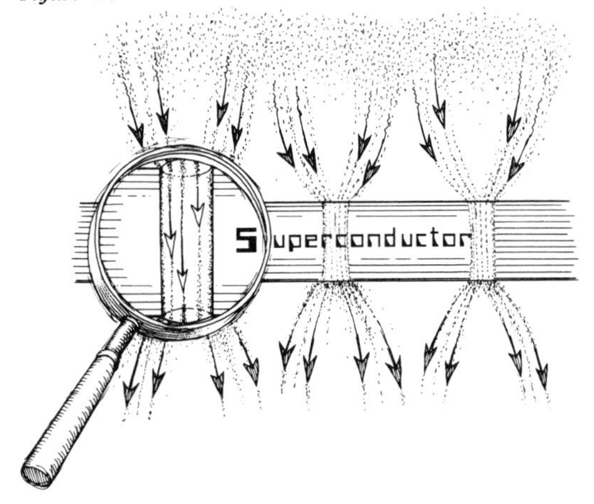

Soliton vortices are also found in superfluids, fluids that can flow without creating turbulence. In this case what forms are not vortices of magnetic flux but long thin cylinders or strings of rotating superfluid that create a curious texture in the superfluid state. Some scientists believe that soliton vortices or "strings" formed in the seconds after the big bang and acted as quantum objects around which matter gathered into galaxies and star clusters.

Another type of soliton penetration, called "self-induced transparency," shows what can happen when light and matter engage in nonlinear interactions.

Whereas crystals such as diamond, quartz, and rock salt are transparent to light, other solids reflect and absorb all the light that falls on them. In these absorbing systems, any light energy that manages to penetrate into the solid is immediately absorbed by its atoms. This absorbed energy then leaks away in the form of atomic vibrations, in other words, heat. So the only effect of trying to force light through an opaque substance is to heat its surface.

However, if the light falling on the solid becomes particularly intense, as in a very high energy burst from a laser, then the solid becomes transparent and the light pulse passes through unabsorbed.

What's the reason for this hat-trick effect? With a sharp laser burst, all atoms in the lattice are pumped up into an excited state. These excited atoms interact nonlinearly with the light so that the two are momentarily fused to form a whole system which, along its wave front, operates collectively. The soliton that passes right through the previously opaque system is not strictly light, neither is it atomic excitation. Rather it is a complex, nonlinear combination of both, a new form of being which theoreticians call a "polariton."

Soliton tunneling also has a role in harnessing thermonuclear energy. Our current form of nuclear energy, fission, makes use of energy that is released when the uranium nucleus breaks apart. In contrast, fusion involves forcing nuclei together rather than fragmenting them.

In a fusion reactor, nuclei of hydrogen or one of its isotopes are heated to such a great temperature that their velocities are sufficient to bind them together when they collide. The collision produces helium and the release of a strong burst of energy. Achieving nuclear fusion requires a combination of very high temperature plasmas—a plasma is a "sea" of freely moving nuclei—and some method of containing the plasma when it reaches a temperature of several million degrees.

While scientists are putting considerable effort and ingenuity into solving the problem of plasma confinement, they are also concerned about heating up the hydrogen plasma to the necessary temperature. One approach is to direct radio waves into the interior of the plasma. The problem is that while these waves heat up the outer regions of the plasma, they can't penetrate far enough into the center where the high temperatures are needed. This is where tunneling solitons come in.

In computer calculations, scientists have discovered that solitons can behave in a strange way when faced with a barrier. Normal linear waves like radio waves are reflected at a plasma barrier, with only a small percentage getting through between the barrier's atoms. With a barrier requiring large amounts of energy to push through, most of the linear radiation is reflected back and very little is actually transmitted. But when nonlinear effects are present, solitons can be generated that will tunnel right through the barrier and come out on the other side without any loss at all. Some scientists believe they will be able to make radio frequency solitons that will tunnel directly into the interior of a plasma to heat it.

BOILING THE UNIVERSE AWAY

Solitons also exist at the smallest levels of nature. Researchers have noted that the results of computer experiments in which solitons collide and interact look suspiciously like the results of experiments carried out using elementary particle accelerators. For instance, the solution to one soliton equation involves what are known as kinks and antikinks. When two kink solitons collide they repel each other, as do two antikinks. However, a kink and an antikink attract. Kinks and antikinks in this respect are identical to oppositely charged elementary particles.

Thinking of elementary particles in terms of solitons has been a hot topic in elementary particle physics. Theorists applying the soliton idea to quantum theory have also come upon something they call the "vacuum bubble instanton." It may be the universe's most lethal and exotic object.

Not only does this quantum nonlinear object strain the imagination, its implications are awesome. Since the instanton comes from merging solitons with a particularly abstract facet of quantum field theory, its story must be told in a series of images and illustrations.

Imagine you are a quantum of energy walking in the mountains and, at the end of a long

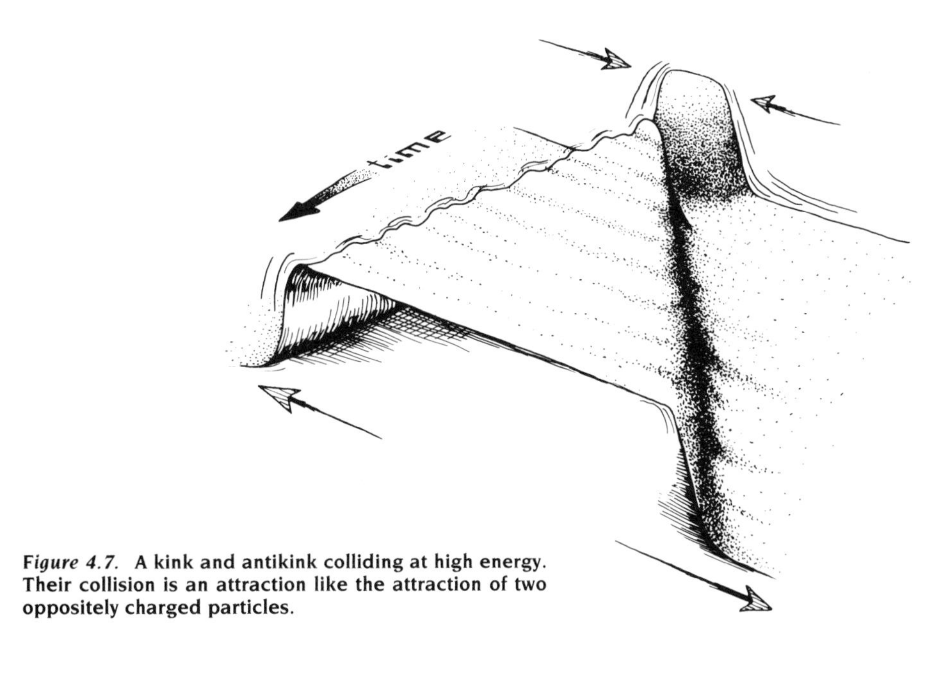

Figure 4.7. A kink and antikink colliding at high energy. Their collision is an attraction like the attraction of two oppositely charged particles.

day, decide to descend into the valley. At the bottom there is nowhere left to go—everywhere else is *up*. Your potential energy is at its minimum so there is no way of lowering it. The valley in which you rest is what physicists call your ground state.

But now suppose you learn that on the other side of the mountain is a deeper valley. Suddenly you realize that your potential energy is in fact not at a minimum, for that other valley is lower.

Of course the only way to get to that valley would be to climb; in other words, the only way to go further *down* would be first to go *up*. But you're a tired little quantum at the end of the day and you don't have the energy for any more climbing.

Now imagine you're the universe. Quantum field theory pictures elementary particles as excitations that arise out of a ground state of the field—also called the vacuum state of the universe. When a little extra energy is added to this vacuum state, elementary particles are created; when that energy leaves, only the vacuum state remains. From the vacuum state there is nowhere left to lose any more energy. It's the bottom of the valley.

But what if on the other side of the mountains of energy that surround the universe there were *another* vacuum state, a potential universe with lower energy? What then? Of course, there is no way of reaching this lower vacuum state from where you are. The energy mountains are too high. In one sense, therefore, your universe would be totally stable, for it can't journey over the mountains to fall down into the deeper vacuum state on the other side. But in an absolute sense, your universe would be unstable because its energy remains high relative to that other ground state. You still have somewhere to fall.

Now we've already met several situations in which solitons are able to tunnel through an energy barrier from one system to the next. Quantum theory also allows this tunneling. Solitons can tunnel from the exterior to the interior of a plasma. Is it possible for solitons formed out of the vacuum state to tunnel from one vacuum to the next? The theoretical model proposed requires another image.

If you heat water to 100° C, as it reaches this temperature small bubbles appear at the bottom of the pan and expand upward. The water has begun to boil. The first bubbles that show up form around tiny dust particles in the water or at cracks and imperfections in the surface of the pan. But if the water is absolutely pure and free from dust and the vessel is totally smooth, "superheated water" occurs, water with too much energy.

Though it is heated several degrees above 100°C, the superheated water looks quite normal; no bubbles reach its surface. But it's unstable. The addition of specks of dust will suddenly provide nuclei for the formation of bubbles and a violent boiling will result.

To complete the analogy, could it be that the ground or vacuum state of our universe is like superheated water—it appears stable enough, but only a single nucleation is required to begin the whole thing "boiling" in a violent outburst of elementary particles? The theory of what physicists call "vacuum bubble instantons" is that a soliton bubble could tunnel across from one ground state or universe to another. This hypothetical bubble becomes quite a bizarre entity, because while its surface belongs to our universe, its interior is alien—it contains the lower vacuum state of another universe.

Theory suggests that if such a bubble arose in our universe it would appear as a violent disturbance expanding outward at the speed of light, causing an exploding froth of energy.

P. H. Frampton of the University of California thinks that a doomsday machine could be devised to produce a single instanton from the intersection of extremely high energy laser pulses. Created with the size of a single elementary particle, after only one second, the instanton would have expanded 300,000 kilometers and would contain within it a "steam" of elementary particles. The result would be like dropping a speck of dirt into superheated water. Our universe would begin to boil.

It is to be hoped that the soliton instanton is purely hypothetical. But other solitons are certainly real. By this point the reader may have wondered, what happens to all those real-world solitons? Where do they go?

The answer is that eventually they die. Though they arise somewhat magically and with a seeming magic resist the forces of dispersion, in time their energy becomes dissipated. Water, for example, has an internal friction or viscosity that acts slowly to erode its form. In time the solitons return to the chaos out of which they arose. Time is the ultimate solvent, we might say.

And it is now to the mysteries of time that we turn for clues to the riddle of how it is that such phenomena as solitons come into being.

Figure 4.8. A satellite photograph of parallel soliton waves about 100 miles apart. Where are all these solitons going?

Chapter 3

The Yellow Emperor said, "Do-Nothing-Say-Nothing is the one who is truly right—because he doesn't know. Wild-and-Witless appears to be so—because he forgets. But you and I in the end are nowhere near it—because we know." Wild-and-Witless heard of the incident and concluded that the Yellow Emperor knew what he was talking about.

CHUANG TZU

CHAOS'S CONNOISSEUR

Soliton behavior seems amazing, but in the view of Ilya Prigogine the sudden appearance of order out of chaos is the rule rather than the exception. A cause for amazement, he insists, is all around us.

Everywhere on the campus of the University of Texas at Austin buildings are taking form. Pile drivers chip into the mocha-colored limestone laid down 100 million years ago when this part of Texas was the continental embankment. In the blocks making up the walls of the modern buildings around the campus are the imprints of ancient seashells. Near one of the construction sites, opposite the Texas Memorial Museum, a slab of the limestone has frozen the footprints and tail-swipes of a sauropod dinosaur.

Surrounded by these images of change and time is a sleek high-rise that houses the physics department. Here in a corner office on the seventh floor, the 1977 Nobel laureate in chemistry ponders the details of his theory that time is the linchpin of creation.

While others are only fairly recently discovering an order to the way systems fall apart into chaos, Prigogine, like an ancient Argonaut, has been thirty years in quest of the secret by which chaos gives birth to order. Along with Poincaré, he is perhaps chaos's prototypical connoisseur.

Asked for the background that led him to his revolutionary track, he says, "You know, I believe in the role of chance and randomness even in life. Therefore there is no logical way of taking one track or another." Born in 1917, at a turbulent moment in Russian history, he was four when his family joined the tide of refugees who emigrated from Russia after the revolution. The family wandered through Europe before finally settling in Belgium in 1929.

He declares that his first passions in schoolwork were history, archeology, and art, "more the human sciences than the hard sciences." But by chance, "it was because of the prewar circumstances in Belgium that I decided to go into sciences at the university." Prigogine recalls that his interest in the humanities meant that "it was very natural for me to become interested in the question of time" and that he was surprised at how little science had to say on the subject. But, again, by chance or destiny, chaos or order, his parents had brought him to a city that was a center for research into thermodynamics, the one scientific field which has attempted to explore the actual meaning of time.

THE OPTIMIST'S AND THE PESSIMIST'S TIME

Thermodynamics, the study of heat transfer and the exchanges of energy and work, is an extremely useful science to engineers but also extremely complex. People are usually familiar with themodynamics from having heard about its famous second law, which predicts that the universe is running down and will eventually succumb to a heat death, or entropy.

First formulated by the German scientist Rudolf Clausius, the second law introduced time and history into a universe which Newton and classical physicists had pictured as eternal. Because the equations of Newtonian mechanics are "time reversible," physicists formed the conviction that at the basic level of matter there is no direction to time. The popular modern illustration of this idea involves the motion picture. If a film of atomic collisions were run backward or forward, we wouldn't know the difference. In the atomic world, time has no preferred direction. Time reversibility is also true of quantum mechanical equations, at least in its conventional interpretation. In fact, the principle of reversibility has survived several revolutions in physics and is a concept so strongly entrenched that Einstein wrote to the widow of a close friend, physicist Michele Besso:

> Michele has left this strange world just before me. This is of no importance. For us convinced physicists the distinction between past, present and future is an illusion, although a persistent one.

The science of thermodynamics, however, discovered a world enthralled by time. Thermodynamically, things go in only one direction. Time is irreversible; it has an arrow. Einstein's friend Besso was born, grew old, and died. His life could never stand still or go backward in time. A car disintegrates into a pile of rust; a pile of rust doesn't reconstitute itself into a car. With the discovery of thermodynamics physicists focused on what might be called a pessimistic time, the time of decay and dissolution.

This aspect of time was fascinating to the young Prigogine, but he was also drawn by the appearance of time in its more optimistic form, evolution. He recalls: "I was very influenced in those years by the beautiful book by Erwin Schrödinger, *What Is Life?* At the end of this book he was asking, where is the organization of life coming from? How does it happen that life is reproducible? That there's some stability in life? Schrödinger said, 'Well, I didn't know. It may be that life has a way of working like a frictionless pendulum.' But I had another idea 40 years ago. My idea was just the opposite. I thought that somehow it is because of the friction and exchanges of energy with the outside world that structure may arise."

Spurred by this intuition, Prigogine pursued his studies at the University of Brussels under Théophile de Donder, one of the few scientists researching what is known as nonequilibrium thermodynamics. Equilibrium is the state of maximum entropy where molecules are paralyzed or move around at random. It is that thingless soup Rudolf Clausius said the universe is heading toward. Learning the laws of nonequilibrium states was a major discovery.

If you take two boxes connected by an opening and put nitrogen in one side and hydrogen in the other, eventually the two will mix so thoroughly that there is essentially no difference in the concentrations of each gas between the two boxes. Scientists say the system has gone to equilibrium and maximum entropy. However, if you heat the two boxes to slightly different temperatures, the gases will mix, but not uniformly. There will be more hydrogen in one side and more nitrogen in the other. The flow of heat has produced some order, that is, it has produced a near-to-equilibrium system.

Near-to-equilibrium can be compared to an energy well in which the system loses heat as fast as it gains heat. The well acts as a point attractor. Prigogine soon realized that even close-to-equilibrium systems have no real sense of time because they keep returning to their attractor. He compares such systems to sleepwalkers or hypnotized persons who have no past. The secret of time Prigogine was looking for was not here.

So after studying near-equilibrium systems for a while, he began investigating what happens in far-from-equilibrium situations—situations undergoing a great deal of energy input from outside. It was here that Prigogine discovered "order out of chaos," and the heartland of time.

Prigogine uses the word *chaos* in two distinct, though sometimes interchangeable ways. There is the passive chaos of equilibrium and maximum entropy, where the elements are so intimately mixed that no organization exists. This is the "equilibrium thermal chaos" of the eventual lukewarm universe predicted by Clausius. But the second kind of chaos is active, hot, and energetic—a "far-from-equilibrium turbulent chaos." This is the chaos that attracted the attention of Feigenbaum, Lorenz, May, Ford, and others we have discussed on the other side of the mirror. Prigogine was one of the first contemporary scientists to discern that in this far-from-equilibrium chaos strange things may occur. He discovered that in far-from-equilibrium states, systems don't just break down, new systems emerge.

Picture a pipe pouring oil into a large basin at an industrial plant. The oil flows smoothly, making a dimple where it enters the surface of the oil in the basin. Now, suppose somebody opens the tap so that more oil flows through the pipe. The first effect of the new spurt of oil is increased turbulence and fluctuations. These fluctuations increase randomly, apparently following a route to total chaos until they reach a bifurcation point. There, at a critical juncture, one of the many fluctuations becomes amplified and spreads, influencing and dominating the system. A pattern of whirlpools forms. Order has sprung out of chaos. These whirlpools remain stable as long as the flow from the pipe is kept up. Even if the flow increases or decreases a little, the stability of the whirlpool pattern remains. Too much change in either direction, however, creates a new chaotic situation and new arrangements of order.

One of Prigogine's favorite illustrations of this order out of chaos is the Bénard instability that we looked at on the other side of the mirror (see pages 50–51). On that side, we considered the route by which the convection cells dissolve into chaos. On this side, however, we consider the way Bénard cells turn chaos into order.

If a pan of liquid is heated so that the lower surface becomes hotter than the upper surface, heat at first travels from lower to upper by conduction. The flow in the liquid is regular and smooth. This is a near-equilibrium situation. However, as the heating continues, the difference in temperature between the two layers grows, a far-from-equilibrium state is reached, and gravity begins to pull more strongly on the upper layer, which is cooler and therefore more dense. Whorls and eddies appear throughout the liquid, becoming increasingly turbulent until the system verges on complete disorder. The critical bifurcation point is reached when the heat can't disperse fast enough without the aid of large-scale convection currents. At this point the system shifts out of its chaotic state, and the previously disordered whorls transform into a lattice of hexagonal currents, the Bénard cells.

Turn the heat up further and the Bénard cells dissolve into chaos.

In his book, *Order Out of Chaos*, written with Isabelle Stengers, Prigogine says that "in chemistry the relation between order and chaos appears highly complex: successive regimes of ordered (oscillatory) situations follow regimes of chaotic behavior." He notes that Bénard cells are a "spectacular phenomenon" produced by millions and millions of molecules suddenly moving coherently.

Clearly a property of far-from-equilibrium chaos is that it contains the possibility of self-

A

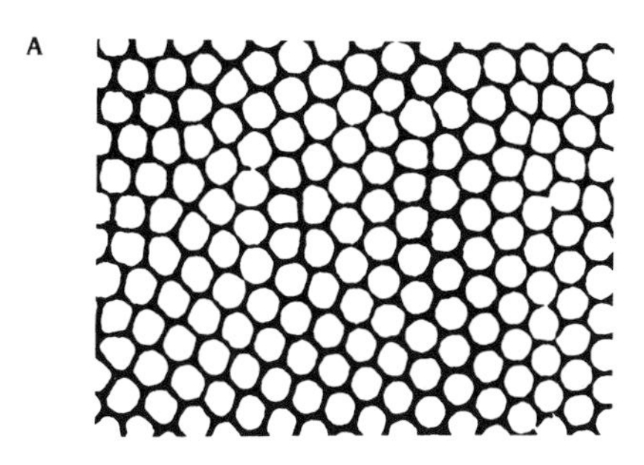

B

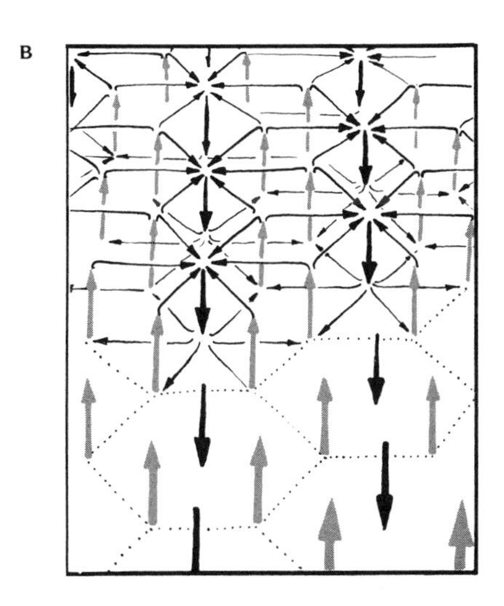

C

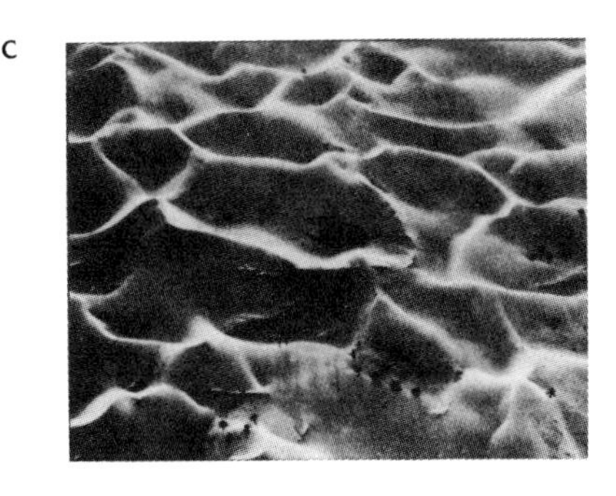

Figure 3.1. **(A) The hexagonal pattern of Bénard cells at the bottom of a pan of heated water. (B) Scientists think the spherical shell of the atmosphere, possibly the whole atmosphere, might be a sea of seething Bénard cells. (C) An aerial photograph of the Sahara Desert shows prints left by this atmospheric Bénard sea. These prints of the atmosphere's convection vortices also show up in snowfields and icebergs.**

organization. Another striking example of self-organization has been found in a whole group of chemical reactions. If the concentration of one of the reagents is increased to a critical point, the reaction undergoes a transformation in which chemical concentrations begin to fluctuate regularly like a chemical clock. Prigogine and Stengers comment in their book:

> Let us pause a moment to emphasize how unexpected such a phenomenon is. Suppose we have two kinds of molecules, "red" and "blue." Because of the chaotic motion of the molecules, we would expect that at a given moment we would have more red molecules, say, in the left part of a vessel. Then a bit later more blue molecules would appear, and so on. The vessel would appear to us as "violet," with occasional irregular flashes of red or blue. However, this is *not* what happens with a chemical clock; here the system is all blue, then it abruptly changes its color to red, then again to blue. Because all these changes occur at *regular* time intervals, we have a coherent process.
>
> Such a degree of order stemming from the activity of billions of molecules seems incredible, and indeed, if chemical clocks had not been observed, no one would believe that such a process is possible. To change color all at once, molecules must have a way to "communicate." The system has to act as a whole.

It is, they say, as if each molecule were "informed" about the overall state of the system. Prigogine is not anthropomorphizing when he talks this way. For him the idea of communication and information is intimately tied up with how random behavior leads to a complex coupling of feedback and spontaneous order. Take as an example how termites make nests.

There is no central termite bureaucracy directing the work. At first termites roam around randomly picking up dollops of earth and transporting them from one place to another. As they do this, they impregnate their bundles with a drop of a chemical that attracts other termites. Randomly, higher concentrations form in some area, which then becomes the focus for other termites and their packets of dirt. Pillars appear and the activity of the termites becomes correlated until the nest is built.

Closer to home, we've all had the experience of finding ourselves involved in such correlations. Driving between rush hours on the thruway, we're only minimally affected by other vehicles. But toward 4 o'clock, traffic becomes heavier and we begin to react and interact with the other drivers. At a certain critical point we begin to be "driven" by the total traffic pattern. The traffic has become a self-organizing system.

Another self-organization that emerges out of chaotic fluctuation involves certain amoebae called slime molds. Slime molds (*Figure 3.2*) spend part of their lives as single cells but when deprived of food send out a chemical pulse which signals other amoebae. Thousands of these amoebae aggregate randomly until their fluctuations reach a critical point, at which time they self-organize to form a cohesive entity capable of moving across a forest floor. Finally, in a new location, the mold develops a stalk and fruiting body which shoots out spores from which new, individual amoebae are born. The slime mold embodies both individual and collective behavior, each aspect enfolded in the other.

As these examples illustrate, Prigogine and his colleagues see self-organizing structures emerging everywhere: in biology, in vortices, in the growth of cities and political movements, in the evolution of stars. He calls instances of disequilibrium and self-organization "dissipative structures."

The name derives from the fact that in order to evolve and maintain their shape, cities and vortices and slime molds use up energy and matter. They are open systems, taking in en-

Figure 3.2

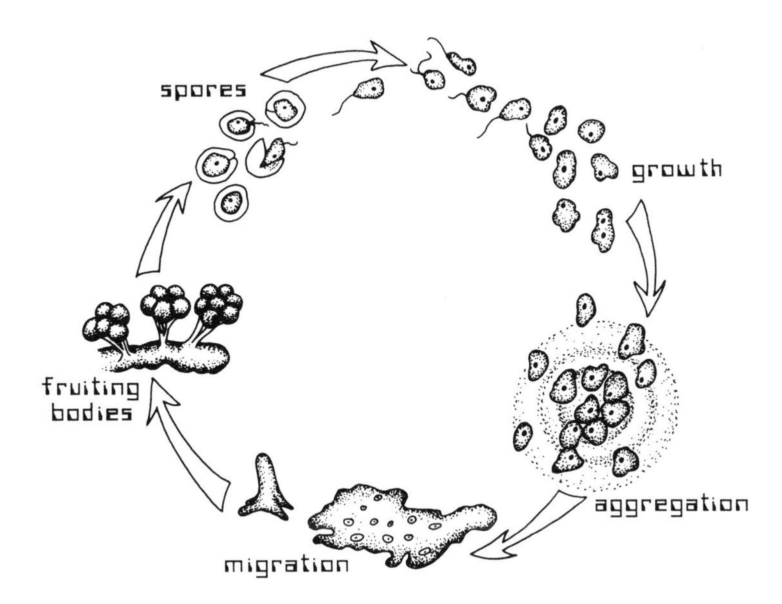

ergy from the outside and producing entropy (waste, randomized energy) which they dissipate into the surrounding environment. Of course one system's entropy may be another system's food; consider the dung beetle, for example, or the mitochondria in our own cells which transform wastes from fermented food molecules into ATP, a molecule that in fact stores energy. The second law (that entropy overall always increases) is not violated by the appearance of these systems, any more than gravity is violated by an orbiting moon. As a moon takes advantage of gravity to stay in orbit, so dissipative structures take advantage of entropy.

The name *dissipative structure* expresses a paradox central to Prigogine's vision. Dissipation suggests chaos and falling apart; structure is its opposite. Dissipative structures are systems capable of maintaining their identity only by remaining continually open to the flux and flow of their environment. Consider the solitons we discovered in the previous chapter. Solitons, like the wave of translation and the candle flame, are also dissipative structures, arising out of a far-from-equilibrium flux and riding upon it.

RADICAL NEW PROPERTIES

It took Prigogine a long time to make breakthroughs in understanding far-from-equilibrium, dissipative structures. "I was in a sense a prisoner of the *linear* nonequilibrium theory," he says, because the near-to-equilibrium systems he studied with de Donder were modeled mathematically by linear approximations. Dissipative structures are creatures of a nonlinear world and at the time he was studying them, there wasn't much scientific interest in nonlinearity.

"Today this seems to be a very, very simple thing, a nearly trivial thing. It's a law now that in the nonlinear range, far-from-equilibrium gives rise to structure, brings order out of chaos. In far-from-equilibrium, matter has radical new properties."

What are these radical new properties enabling self-organization? How does dissipative structure construct itself out of a chaotic background—organizing space, giving an inexorable direction to time?

In an ordinary reaction, molecules of two chemicals engage in random motion. In some collisions the molecules chance to have the correct energy and orientation and so stick together to form a new molecule, the reaction's "product." The collisions continue until all the starting molecules are combined into the product. The system ends up with homogenous, unstructured mixtures of chemicals.

But in some reactions, one type of molecule can't be made unless it finds itself in the presence of another of its own type. Such a chemical becomes its own catalyst. It iterates. Chemists call these reactions "autocatalysis," "cross catalysis," and "autoinhibition" because they involve processes in which the products of some steps feed back into their own production or inhibition. Such chemical iterations lead to chemical systems which exhibit everything from equilibrium and limit cycles to period doubling, chaos, intermittency, and self-organization. The systems structure space by grouping the reacting molecules into orderly patterns of a certain size, and they mark time by evolving and changing constantly. They are never quite the same even when they are maintaining the same basic organization.

One of the most colorful of these reactions is a purely chemical critter called Belousov-Zhabotinsky (*Figures 3.3, 3.4*).

Scientists have recently been able to replicate the growth of structure in Belousov-Zhabotinsky on the computer using iterative nonlinear equations. In real life the reaction appears when malonic acid, bromate, and cerium ions are mixed together in a shallow dish of sulfuric acid. The right concentrations and the right temperature are necessary for the scrolls to come out, and the reaction first goes through a period of chaos. The form that

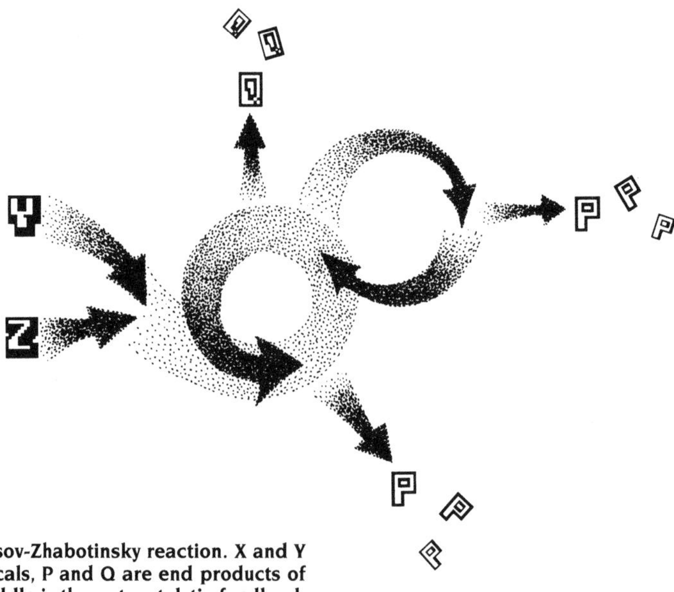

Figure 3.3. **The Belousov-Zhabotinsky reaction. X and Y are the starting chemicals, P and Q are end products of the reaction. In the middle is the autocatalytic feedback iteration that sustains the reaction.**

Figure 3.4. Overview in a dish as the Belousov-Zhabotinsky reaction unfolds. The nonlinear iteration of this reaction means that the initially random or chaotic motions of molecules in solution spontaneously give rise to structures in space and time. The slightest fluctuation in one part of the solution may become magnified. Speaking metaphorically, if there is a chance concentration of "red" molecules in one region, these "red" molecules act to catalyze or assist the production of more "red" molecules. Thus the red chemical builds up in one region, while the blue chemicals prevail in a nearby region. In this way, large-scale structuring of the different chemicals occurs. Order emerges out of chaos thanks to the energy that is constantly being supplied by the chemical reaction.

emerges has complex levels of detail and can self-reproduce its structure much like something alive.

It would have taken nature far longer than the age of the universe to come up with a self-reproducing sequence of amino acids like DNA if the process had been left purely to chance. However, the self-organizing ability of the kinds of chemical reactions like Belousov-Zhabotinsky that existed on the early earth suggests that rather than being merely a chance occurrence, the order we call life is a variation on a very old theme.

Astronomers investigating the way disk galaxies are formed have suggested how old a theme it is. They've concluded that the same autocatalytic (iterative) model which works to produce the scrolls of the Belousov-Zhabotinsky reaction applies to the scroll formation of these ancient structures millions of light-years in size.

But the scroll-like birth of order has its other face—the growth of chaos. The normal contractions of the heart expand from a trigger point in circular wave front across the heart's surface. If this wave is broken anywhere, it gives rise to complex spiral disturbances which are self-replicating and very resilient. The heart's response to these electrical spirals leads to the fractal forms and period doubling of heart failure. A similar effect is thought to cause some kinds of epileptic seizures.

Thus the same processes of bifurcation, amplification, and coupling may lead to one side of the mirror or to the other.

Between four and six thousand years ago the ancient peoples in Europe built stone circles and decorated them with interlocking

Figure 3.5. An astronomical version of the Belousov-Zhabotinsky reaction?

Figure 3.6. **Spirals adorn a Stone Age tomb in Ireland.**

scroll loops. Similar motifs appear all over the world. The psychologist Carl Jung said such images are archetypes or universal structures in the collective unconscious of humankind. Could such a collective wisdom perhaps be expressing its intuitions of the wholeness within nature, the order and simplicity, chance and predictability that lie in the interlocking and unfolding of things?

BIFURCATION: WINDOW OF THE FORKING PATHS

Like a momentary window into the whole, the amplification of bifurcations leads to order, or to chaos. In Prigogine's scheme of things, *bifurcation*—a word which means the place of branching or forking—is an essential concept. A bifurcation in a system is a vital instant when something as small as a single photon of energy, a slight fluctuation in external temperature, a change in density, or the flapping of a butterfly's wings in Hong Kong is swelled by iteration to a size so great that a fork is created and the system takes off in a new direction. Over the course of time, cascades of bifurcation points either cause a system to fragment itself (period doubling) toward chaos, or to stabilize a new behavior through a series of feedback loops (such as autocatalysis, cross catalysis, and autoinhibition) that couple the new change to its environment.

Once stabilized by its feedback, a system that has passed through a bifurcation may resist further changes for millions of years until some critical new perturbation amplifies the feedback and creates a new bifurcation point.

At its bifurcation points, the system undergoing a flux is, in effect, being offered a "choice" of orders. The internal feedback of some of the choices is so complex that there is a virtual infinity of degrees of freedom. In other words the order of the choice is so high that it is chaos. Other bifurcation points offer choices where the coupling feedback produces fewer degrees of freedom. These choices may make the system appear simple and regular. But this is deceptive, for the feedback in simple-appearing orders like the soliton wave is also unanalyzably complex.

The net result of bifurcations in the evolution of living cells has been to create organic chemical reactions that are intricately and stably woven into the cell's environment. This weaving of feedback loops is what Prigogine means by "communication." Through such communication the system holds itself intact.

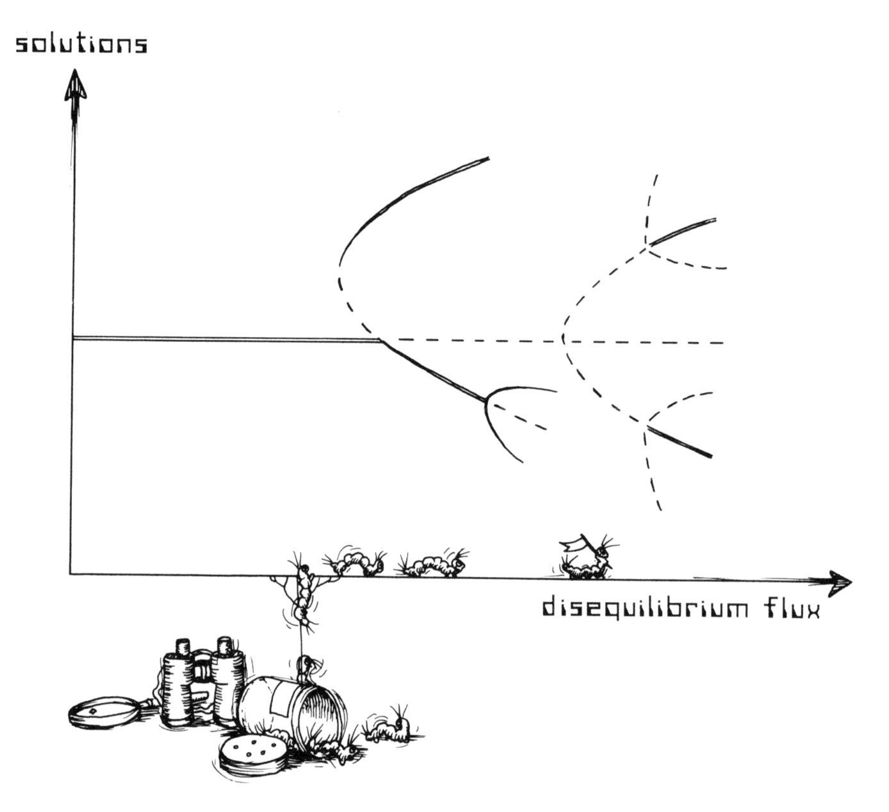

Figure 3.7. **The broken lines in this diagram indicate unstable states (chaos). The solid lines are stable or "steady state" solutions that the system can maintain in the flux. The diagram could represent many different kinds of systems, chemical or biological. As the dotted lines show, when the flux is increased, the system passes through instabilities where it is faced with choices. Most of these lead to chaos, some to order. The ones that lead to order are stabilized by coupling iterations, creating an interlocked net of feedback.**

Bifurcation points are the milestones in the system's evolution; they crystallize the system's history. The historical record of our own bifurcations is etched in the shape of our lungs with its Fibonacci/fractal shifts in scales (see page 107). A document of our past bifurcations appears in our embryos as we pass through stages where we resemble fish, then amphibians and reptiles.

Enfolded in all the shapes and processes that make us unique—in the chemical reactions of our cells and the shape of our nerve nets—are thousands upon thousands of bifurcation points constituting a living chronology of the choices by which we evolved as a system from the primordial single cell to our present form.

At each bifurcation point in our system's past, a flux occurred in which many futures existed. Through the system's iteration and amplification, one future was chosen and the other possibilities vanished forever. Thus our bifurcation points constitute a map of the irreversibility of time.

Time is inexorable, and yet in bifurcations the past is continually recycled, held timeless in a sense—for by stabilizing through feedback the bifurcation path it takes, a system embodies the exact conditions of the environment at the moment the bifurcation occurred. A vestige of the primal sea remains in the chemical reactions linking the mitochondria in our cells to the cytoplasm surrounding them; the landscape of the age of reptiles lurks in the structure of our brain's reticular activating system, which governs our level of alertness.

Thus the dynamics of bifurcations reveal that time is irreversible yet recapitulant. They also reveal that time's movement is immeasurable. Each decision made at a branch point involves an amplification of something small. Though causality operates at every instant, branching takes place unpredictably.

Prigogine says, "This mixture of necessity and chance constitutes the history of the system." It also constitutes the system's creativity. The ability of a system to amplify a small change is a creative lever.

Biological systems remain stable by damping most small effects except in those areas of behavior where a high degree of flexibility and creativity is required. Here the system remains highly sensitive to influx, close to a state of chaos. A single honeybee entering a hive of thousands of interacting colleagues can, by doing a little dance indicating the location of pollen-rich flowers, launch the entire hive into the air.

Systems are also highly sensitive near those places that are the crystallized "memory" of past bifurcations. Nations have usually evolved through bifurcations involving intense conflict. Consequently, they are highly sensitive to certain kinds of information, which recreate those bifurcations. A single headline in a newspaper can mobilize an entire nation for war.

The idea of birfurcation sensitivity is also being marshaled to explain the curious phenomenon of chirality.

Chirality means handedness, the fact that we live in an asymmetrical world. The patterns on seashells whorl in one direction more than another. The important molecules of life are mostly left-handed. In the laboratory it's possible to produce molecules with an equal probability of left- and right-handedness; in fact, it's difficult to get laboratory chemical reactions to have asymmetry unless you seed in some handedness from the outside. But this is not the case in nature. Louis Pasteur, one of the first to study the problem, concluded that there must be a basic asymmetry in nature, but he was never able to find the origin for this asymmetry. There have been various theories to explain chirality since Pasteur, none of them entirely satisfactory.

Recently members of Prigogine's group came up with a solution which they published in the scientific journal *Nature*. In the 1970s physicists were surprised to discover that the world of atomic particles is itself not quite symmetrical. When electrons are sprung loose from the atom, they come out spinning clockwise or counterclockwise. Physicists now talk about God as left-handed. However, the energies involved in a choice of elementary particle handedness are minuscule when compared to the energies involved in the molecules of life. Scientists were convinced left-handedness at the level of the elementary particles could have nothing whatever to do with the left-handedness of biological molecules.

But, as we have seen, in far-from-equilibrium states very small effects become magnified. For example, the extremely small gravity difference across a few centimeters of liquid would normally be negligible. In the case of the Bénard instability, however, the far-from-equilibrium turbulence hugely magnifies this gravitational effect and results in the hexagonal Bénard pattern.

Prigogine's colleague, D. K. Konepudi, believes that something similar happens with the small preferential spin of electrons. In the disequilibrium chaos that gives birth to new molecules, a dissipative system may quickly amplify the very small energy difference in spin, projecting God's subatomic left-handedness up to the level of the organic molecule.

WHAT DIRECTION FOR TIME?

Despite the hypothetical connection between the spin of subatomic particles and the formation of handedness in large-scale molecules, most scientists continue to believe that there exists an essential division between the small-scale quantum and large-scale "classical" Newtonian realms. The nineteenth century's discovery of irreversible time in its optimistic and pessimistic forms—in entropy and evolution—did nothing to dissuade physicists from their conviction that at the most basic levels of matter, time is reversible and that the irreversibility we see around us is—as Einstein said to Besso's widow—a kind of illusion. This conviction stems from the time reversibility of the linear equations describing the movement of atoms and elementary particles. In the 1870s, Boltzmann had overcome the apparent paradox between atomic and classical scales by arguing that the answer lies in the way atoms shuffle themselves, making the restoration of an initial order increasingly improbable. He argued that irreversibility enters into the world because the reversible collisions that systems undergo are so complex that, like sleepwalkers, atoms "forget their initial conditions" and become disordered. Boltzmann's brilliant solution linking the classical Newtonian science of gravitation to the thermodynamic science of change led to the invention of the scientific field of statistical mechanics.

As a result of Boltzmann's efforts, thermodynamics became the rage in the latter part of the nineteenth century. It gave important fuel to the logical positivist form of reductionism. The positivists believed that all phenomena could be reduced to a description involving mechanical dynamics, that is, bodies in motion. Thermodynamics and the transformation of energy were assumed to involve this sort of mechanical dynamics.

Even Freud was deeply influenced by the positivist reductionist approach and Freudian ideas were originally developed in thermodynamic terms. Freud spoke of psychoanalysis as "a dynamic conception which reduces mental life to the interplay of reciprocally urging and checking forces."

All was not totally rosy for thermodynamical reductionism, however. Poincaré complained that Boltzmann's solution to the reversibility-irreversibility dilemma was a conjuring trick that failed to solve the problem at a truly fundamental level. Boltzmann perhaps agreed; in despair over what he regarded as a failure of his explanation, he committed suicide.

Prigogine says that at first he accepted Boltzmann's solution and the belief that the fundamental laws of physics are time reversible. "I believed, as everybody did, that there is irreversibility but that it must come from approximations which we are forced to make on the basic time-reversible rules; it comes from ignorance, from our approximations." However, "far-from-equilibrium studies led me to the conviction that this cannot be the right point of view. Irreversibility has a *constructive* role. It makes form. It makes human beings. How could our mere ignorance about the initial conditions be the reason for this? Our ignorance cannot be the reason we exist."

It can't be, he insists, that if we could only increase our knowledge, make a computer powerful enough so that we could write equations for the movement of all the reversible and probabilistic individual molecules constituting a system, our ignorance would disappear, the illusion of irreversibility would be exposed, and with it life, evolution, death, and time itself would vanish. "That's paradoxical."

Chaos theory now supports him in this position, for, as we've seen, no computer can ever be made large enough to track an irreversible system. Our ignorance is an expression of the holistic fact that in the universe of the dynamical forces which create galaxies and cells, all things are interwoven. That is the true meaning of irreversibility.

Nevertheless, there are those linear equations which tell physicists that, stripped down to the bare atomic particles, reversibility must rule. Prigogine balks at this established "fact." And in the past few years he has been devoting his efforts to mounting a daring assault on the foundations of physics at exactly the period of history when some eminent physicists feel that the time-reversible equations are on the brink of explaining virtually everything we ever wanted to know about how atomic things work.

Against this confidence, Prigogine poses a nagging, commonsense objection that harkens back to Poincaré. Even on the microscopic level, he says, reversibility is the illusion. "You can never make an experiment in which the past and the future are the same for an unstable dynamical system of atomic particles. If we start with particles which have the same velocities, and have collisions, they will end up with random velocities. But we cannot make the reverse experiment. There are no reversible experiments. Therefore, our world is temporally organized." There is always an arrow to time. Prigogine also points out that relativity, which Einstein envisioned as a statement of the reversibility and interchangeability of space and time, has led to the formulation of the big bang theory, which in fact gives the universe an irreversible history. In present-day quantum physics, he contends, irreversibility shows up everywhere.

If the first challenge he offers to his contemporaries is a challenge to reversibility, the second is to the concept of simplicity. Ever since Democritus and Aristotle, scientists have believed that underlying the complexity of our world there must be simple objects and forces. At first scientists thought these simple building blocks were atoms. Later, when atoms were found to have parts, the building blocks became simple particles like the proton and electron. Then, when quantum mechanics led to the unexpected discovery of a staggering "particle zoo" at the subatomic level, physicists devised the grand unified theory and began looking for the single, simple force—the "superforce"—that allegedly gave birth to this maze of small-particle interactions. The superforce has not yet been found, however, and, at least so far, research has discovered that for every simplification there are at least two new complexities. Prigogine says, "The idea of simplicity is falling apart. Any direction you go in there's complexity."

Then how does it all work? Fashioning his revolutionary proposal to unify dynamics and thermodynamics, the microscopic world and macroscopic world, reversibility and irreversibility, being and becoming, Prigogine makes the argument that time is a form of "symmetry breaking."

Scientists conceive of empty space as rotationally symmetrical in that all directions are equivalent. Introduce a magnet like the earth into that space, however, and the symmetry is broken. The magnet singles out north as a special direction and from then on it becomes possible to orient other magnets in space.

In a similar way, Prigogine argues, complex systems break the symmetry that would allow time to go backward as well as forward. Complex systems give a direction to time. How do they do that?

Complex systems—both chaotic and orderly ones—are ultimately unanalyzable, irreducible into parts, because the parts are constantly being folded into each other by iterations and feedback. Therefore, it's an illusion to speak of isolating a single interaction between two particles and to claim that this interaction can go backward in time. Any interaction takes place in the larger system and the system as a whole is constantly changing, bifurcating, iterating. So the system and all its "parts" have a direction in time.

Time thus becomes an expression of the system's holistic interaction, and this interaction extends outward. Every complex system is a changing part of a greater whole, a nesting of larger and larger wholes leading eventually to the most complex dynamical system of all, the system that ultimately encompasses whatever we mean by order and chaos—the universe itself.

Once any complex system appears, Prigogine says, it becomes separated from reversible time by what he calls an "infinite entropy barrier." Processes that run in the reversed time direction become not just astronomically improbable, as Boltzmann had said, but *infinitely* improbable. This can be illustrated by thinking about the ripples spreading out from a stone thrown into a pond. To time-reverse this situation would require coordinating precisely all the infinitesimal disturbances around the edge of the pond so that they move inward, growing in amplitude and finally converging at a single dimple. While the nonlinear coupling of forces needed to produce a soliton wave is staggering, the linear coordinating of forces here would be *infinitely* staggering.

Any ultimate coordinating of events around the pond is made impossible by the fact that all systems are open to the rest of the universe. Nature is bathed in a constant flux of gravity, electricity, and magnetism in addition to small fluctuations in temperature and other forces. Even the movement of distant stars will produce minute changes in the gravitational field experienced on earth. While these fluctuations will be beyond any hope of measurement on earth, nevertheless they will always destroy initial correlations. So even if the correct initial conditions could be set up around the edge of the pond, they would be rapidly obliterated by such subtle contingencies long before the contracting ripples converged in the center. In ideal, isolated systems time may be reversible but in *real* systems the symmetry of time is always broken.

Prigogine believes the symmetry breaking of time occurs at all levels of nature, from the quantum to the elephant to the galaxy. There is both one time and infinite times. Time is the great arrow which couples all systems together and a multitude of arrows which constitute the bifurcations and changes of each individual system. Each of us has his or her own autonomous irreversible arrow but that arrow is intertwined with the irreversible arrow of the universe.

Using this logic, Prigogine revises the big bang theory. He says: "The universe starts with a burst of entropy (chaos) which leaves matter in an organized state. And, after this, the matter is slowly dissipating and creating in this dissipation, as a by-product, cosmological structures, life, and, finally, ourselves. You see, there is so much entropy dissipated that you can use it to build something." Thus the entropy which Clausius saw as mere random soup is for Prigogine an infinitely nourishing soup from which appear dissipative structures. Prigogine revises the classical concept of entropy, or passive chaos, by making it active. Entropy has, he says, "both positive powers and negative powers. The positive powers are used to compensate the negative powers in such a way that the total remains positive."

Prigogine believes classical and quantum dynamics, with their insistence on reversibility and timelessness, are an idealization of nature. As we discovered on the other side of the mirror, a system can never be sealed in a box. The "outside," the whole, always leaks in through the break in the chain of decimals, the "missing information." Thus actual nature is always entropic, turbulent, and irreversible. By seeing irreversibility all the way down to the bottom of things, Prigogine wants to do away with the traditional scientific separation between the large-scale and small-scale universe. "If not rooted in the microscopic world, then where does our world come from? Where does our time come from? . . . It's a very remarkable fact. If you take many of the greatest men in the last two centuries—Bergson, Heidegger, Einstein—they would all say that irreversibility cannot be found through physics. It has to be either found through metaphysics or it's something we add to nature. If you take this point of view, time separates us from the universe. But if you think that irreversibility is a natural phenomenon, then time is no longer separating us from nature."

A moment later he says reflectively, almost wistfully, "If I have to reverse the classical concepts it's not because I want to, but because I'm trying to express what I see about the constructive role of irreversible processes. . . . I have not started my work by saying I want to introduce new concepts."

That may be, but Prigogine's view has not gone down well with his contemporaries. A reviewer of his last book noted that "the closest thing to a consensus about Prigogine's work is that it falls somewhere on the spectrum bounded by responsible science and the Maharishi Mahesh Yogi's Technology of the Unified Field." The late Heinz Pagels charged, "Only Prigogine and a few collaborators hold to these speculations which, in spite of their efforts, continue to live in the twilight zone of scientific credibility."

Pagels, a physicist and the well-known author of *The Cosmic Code*, was an orthodox quantum physicist. His scathing critique may reflect an attitude that physicists sometimes have about chemistry, and Prigogine's principal field is chemistry. Physics is the queen of sciences because it deals with nature's most basic laws. Chemistry studies derivative issues, this line of reasoning goes. But this very attitude is being attacked in Prigogine's third challenge to the scientific establishment. Nature has been traditionally viewed as a hierarchy beginning with atomic structure and ending with complex biological organisms. Each level of scientific description is supposedly built on the preceding, with descriptions at the most fundamental level—physics—having priority. But for Prigogine, nature is not built from the bottom up. It is built by feedback among all levels. Thus his idea of a scientific description of nature "does not suppose any fundamental mode of description; each level of description is implied by another and implies the other. We need a multiplicity of levels that are all connected, none of which may have a claim to preeminence."

This kind of statement has naturally rankled some physicists. Its companion statement is even more provocative. Prigogine believes that the laws of nature, including the laws of physics, are not all "given" at the outset or even logically implied. They evolve the way different species evolve. As things get more complex, bifurcations and amplifications occur and new laws emerge. "How can you speak about the laws of biology if there're no living systems? Planetary motion is something coming in very late."

This is an assertion of nature's creativity. Each level of organization produces something fundamentally new, something that is not present in the constituent elements or "parts" of the previous level. For example, water is not present in a mixture of hydrogen and oxygen. It has a new unity which, in effect, sacrifices the "parts" hydrogen and oxygen. The only way to get these parts back is by destroying the water.

Since no laws or "parts" of the universe are more fundamental than any other, rather than proceeding linearly and hierarchically, Prigogine believes that science must try to sort out and describe the network of laws and processes that join all levels. Nature is to be viewed as a dynamical shifting web, not a mechanical, hierarchical pyramid.

Pagels was a dedicated advocate and participant in work on the grand unification project to discover the superforce underlying all of matter. Small wonder he was disturbed by Prigogine's approach. Prigogine has his own critique of the grand unification project: "Grand unification wants to reach a description of the universe which is unified, but if it's unified then you have no second law of thermodynamics [the law of increasing entropy—time]. The universe isn't an identity, all the particles don't melt down into one. If you have an identity, you don't have an arrow of time, and that arrow exists."

CREATIVE CHAOS

Behind Prigogine's arguments for time smolders the soul of a visionary who believes that in the laws of unpredictability, chaos, and time —not in the mechanical laws of classical dynamics—lies the secret of nature's creativity. He cites as an example of the creativity of chaos and irreversibility their role in the emergence of life.

Konepudi and other scientists in Prigogine's research group are working on experiments that may show how the complicated code on the DNA nucleotide evolved. DNA is a polymer or chain of molecules with recurring links. The problem is, as Prigogine puts it, "How can you put a text on a polymer?" Or more properly, how can you turn a polymer *into* a text? "If you couple a polymer with a limit cycle reaction, the polymer will become ABAB. You will not do very much. But if you couple it with a chaotic reaction, you obtain complicated sequences. You obtain a symbolic dynamics"—in other words, a record of the birth of different bifurcations of dissipative structures, the order in chaos. "This text is information-rich. And because this text is due to irreversibility, there is a way of reading it, which is just what we find in the real nucleotides, which you have to read in one direction." The nucleotide is, therefore, a record of the far-from-equilibrium dynamics (the environment) that acted upon it and it can reproduce those dynamics as it's read. "Therefore you see that chaos is not at all this negative element," Prigogine says triumphantly.

By emphasizing the role of randomness and chaos in the creation of structure, Prigogine conjures up a universe in which objects are less well defined than in classical and even quantum physics. In Prigogine's cosmos the future can't be determined because it is subject to randomness, fluctuation, amplification. Prigogine calls this a new "uncertainty principle."

The famous uncertainty principle formulated for quantum mechanics by Werner Heisenberg stated the impossibility of knowing with total accuracy both the position and the momentum of any subatomic particle. The uncertainty principle introduced the need for probability in the description of particle behavior. Prigogine's new uncertainty principle says that beyond a certain threshold of complexity, systems go in unpredictable directions; they lose their initial conditions and cannot be reversed or recovered. This inability to go backward in time is an "entropy barrier." The discovery of the entropy barrier is similar to Einstein's discovery that human beings and messages can never travel faster than the speed of light, that is, beyond the "light barrier."

Like Heisenberg's uncertainty principle, Prigogine's uncertainty principle is a blow to reductionism. But for Prigogine this way of looking at nature is less a limitation than a recognition of creative possibilities. For example, in talking about the ideas of progress, he and Stengers note that most definitions of progress give "a reassuring representation of nature as an all-powerful and rational calculator, and of a coherent history characterized by global progress. To restore both inertia and the possibility of unanticipated events—that is, restore the open character of history—we must accept its fundamental uncertainty. Here we could use as a symbol the apparently accidental character of the great cretaceous extinction that cleared the path for the development of mammals, a small group of ratlike creatures."

In the end of their book, they describe their irreversibility approach—the approach of chaos—as a trend leading to something new in science: "a kind of 'opacity' as compared to the transparency of classical thought." One is reminded of John Keats's proclamation that in order to be a poet you must be able to live in "doubt and uncertainty." Prigogine is proposing this as a new path for science.

> Is this a defeat for the human mind? This is a difficult question. As scientists, we have no choice; we cannot describe for you the world as we would like to see it, but only as we are able to see it through the combined impact of experimental results and new theoretical concepts. Also, we believe that this new situation reflects the situation we seem to find in our own mental activity. Classical psychology centered around conscious, transparent activity; modern psychology attaches much weight to the opaque functioning of the unconscious. Perhaps this is an image of the basic features of human existence. Remember Oedipus, the lucidity of his mind in front of the sphinx and its opacity and darkness when confronted with his own origins. Perhaps the coming together of our insights about the world around us and the world inside us is a satisfying feature of the recent evolution in science that we have tried to describe.

The reductionist biologist Jacques Monod has characterized the science that developed in the first part of this century, the science of quantum mechanics, and its descendant, grand unification, as defining a universe in which life and human beings are accidents which do not "follow from the laws of physics" but are "compatible with them." As a consequence, science has shown man to be alone and isolated in the cosmos, a "gypsy" who "lives on the boundary of an alien world, a world that is deaf to his music, just as indifferent to his hopes as it is to his suffering or his crimes." The universe is a vast machine, a probabilistic computer in which life and intelligence have only a comparatively low probability.

Prigogine soundly rejects the reductionist view. By focusing on the existence of time in all dimensions of reality, on the pervasive role of chaos in creating spontaneous order, Prigogine thrusts toward what he calls "the reenchantment of nature." He wants to show that as time-bound, spontaneously created beings, we are an integral part of the time-bound, spontaneously organized movement of nature, not a low-probability accident. He also wants to show that what we do makes a difference.

"Freedom and ethics have no place in an automaton. However, once you see that the world is sufficiently complex, then the problem of value becomes different.... What we are doing leads to one of the branches of the bifurcation. Our action is constructing the future." He believes that "since even small fluctuations may grow and change the overall structure, as a result, individual activity is not doomed to insignificance. On the other hand, this is also a threat, since in our universe the security of stable, permanent rules seems gone forever. We are living in a dangerous and uncertain world that inspires no blind confidence."

Renée Weber in her book *Scientists and Sages* places Prigogine with traditional mystics and the new scientific mystics like David Bohm.

But she shows that Prigogine is not a mystic in an easily definable sense. Certainly he seems mystical about chaos. For example, though the word crops up repeatedly in his conversation and writing, he refuses to define it. But he is not a believer in the direct mystical perception of Oneness.

In defining his brand of mysticism it seems appropriate to note that he is a collector of pre-Columbian art filled with earthy and misty forms. His mysticism seems closer to this: to art, or to an ancient science bringing news that the Yellow Emperor would welcome. He carries word that the spell of reductionism was a dream—and the waking time-bound reality around us is an even better dream.

Native Americans have a saying that time is timeless, and that this is a fact which the Indians have always known but the white man has yet to learn. Perhaps Prigogine is one of the first Western scientists to learn it, or relearn it. Then to the timelessness of time he has added another ingredient, also an old one: chaos as the source of structure and life.

Chapter 2

When the Yellow Emperor woke, he was delighted to have found himself.

LIEH-TZU

THE AUTONOMY COLLECTIVE

Prigogine's insights into chaos highlight the difference between a mechanical view of nature and a holistic one. Another way to understand this vast difference is by looking at feedback.

If a machine malfunctions, finding the problem is relatively easy. A link in the cause-and-effect chain of parts has broken. Find the link and make the repair. However, when the human body malfunctions, a doctor may diagnose some particular culprit, but in fact the "cause" for any disturbance of our health is always multiple, since a living organism is made up of a staggering number of feedback loops. Woven into the loops of living structures are the transmutation of food into energy, the contraction of muscles, the regulation of body temperature, the movement of hormones and neurotransmitters, the actions of reflexes such as the dilation of the iris of the eye in sudden darkness, or the quickening of the heart in the presence of danger. Negative feedback loops regulate, positive loops amplify. Myriads of loops are hooked together in such a way that the internal organization of an organism can continuously adjust to the demands of its environment. A machine can be completely disassembled into its parts and put back together so that it runs normally, but this can't be done with a living entity. If a working part of a machine is lost, the machine grinds to a halt. If a working part of an organism is lost, however, the organism may compensate for the missing bit through its feedback loops and go on. Finally, a machine converts fuel into heat and motion, but it does not convert fuel into itself as an organism does through its feedback.

The feedback properties described above, particularly the property of constant self-renewal, give living systems unique characteristics. These characteristics are defined by scientists in the concept of "autopoiesis."

Autopoietic structures lie at the highly sophisticated end of nature's spectrum of "open systems." The spectrum runs from simple self-organizing systems like whirlpools and Jupiter's Red Spot to more complicated chemical dissipative structures like the Belousov-Zhabotinsky reaction and finally to highly complex autopoietic systems such as ourselves. Autopoietic systems are remarkable creatures of paradox. For example, because autopoietic structures are self-renewing, they are highly autonomous, each one having its separate identity, which it continuously maintains. Yet, like other open systems autopoietic structures are also inextricably embedded in and inextricably merged with their environment—which is necessarily a far-from-equilibrium environment of high energy flows involving food, sunlight, available chemicals, and heat. To express the paradox another way: Each autopoietic structure has a unique history, but its history is tied to the history of the larger environment and other autopoietic structures: an interwovenness of time's arrows. Autopoietic structures have definite boundaries, such as a semipermeable membrane, but the boundaries are open and connect the system with almost unimaginable complexity to the world around it.

The fast-action films that have been made of people involved in conversation illustrate the autopoietic paradox. The films show that a subtle dance occurs between speaker and listener, a rhythmic action back and forth, as if precisely choreographed. The viewer of the film seems to be in the presence of the movements of a single organism. The conversation reveals the subtle interconnectedness that underlies all autonomous structures. Similarly, our most private thoughts and feelings arise out of a constant feedback and flow-through of the thoughts and feelings of others who have influenced us. Our individuality is decidedly a part of a collective movement. That movement has feedback at its root.

THE NONLINEAR PLANET

The feedback nature of autopoietic structures should not, perhaps, be surprising, since from the outset, life on earth has been built by feedback interconnectedness. That interconnectedness—it should also not be surprising—had its roots in chaos.

Recall the stunning complexity that took place at the boundary of the Mandelbrot set. Now picture the gyrating shapes of that purely mathematical iteration as a metaphorical view of the chemistry that once bubbled and coagulated on primordial earth.

According to Sherwood Chang of NASA's Ames Research Center, the dissipative structures that led to life on the planet probably began at the chaotic interface between solid, liquid, and gaseous surfaces where there is flux of high energy. Some scientists speculate that at this chaotic nexus, autocatalytic chemical structures like the Belousov-Zhabotinsky reaction constituted a form of protolife and that on early earth many variations of such reactions flourished. Responding to the far-from-equilibrium environment, the descendants of these first autocatalytic, self-referential, self-similar structures linked together to form a larger structure of feedback loops called a hypercycle. One hypercycle structure was RNA.

The emergence of RNA and its important descendant, DNA, were dramatic new steps in the theme of self-similarity born out of chaos. Through RNA and DNA, the hypercycle's ability to iterate and replicate itself became greatly enhanced. Since DNA's copying process also created variations, the interactions not only reproduced the same forms, they produced reams of new ones. The microbes which the RNA hypercycle gave birth to were fantastically adaptable to the harsh conditions of early earth.

The myriad varieties of microbes which first inhabited our planet, and which still inhabit it, adapt by passing around fragments of DNA. A bacterial "strain" can be altered by the simple expedient of reshuffling its genetic code by taking in new bits of DNA or giving away old bits. Employing this method, bacteria transformed the earth. The method allowed teams composed of different strains of bacteria to couple together, the waste products of one strain becoming the food sources of another.

Systems theorist Erich Jantsch once pointed out that if the drive of evolution were simply adaptation, then evolutionary change should have ceased with the bacteria. Bacteria's DNA feedback mechanism makes it possible for them to mutate and adapt to all kinds of adverse conditions with amazing speed. But evolution apparently has other drives, Jantsch proposed, one of which may be, as he put it, the pure "intensification of life." At the next stage of intensification, biological feedback evolved into a radically new form.

There is growing support among scientists for a revolutionary feedback theory of evolution advanced by Boston University microbiologist Lynn Margulis. Margulis believes that the "new kind of cell" which made its appearance 2.2 million years ago to become the basis for the cells of all the multicelled plants and animals that exist today was not the result of a genetic mutation, but of symbiosis. It was not the product of brutal competition for survival of the fittest, but of cooperation. In her book *Microcosmos*, written with her son Dorion Sagan, she says:

> Competition in which the strong wins has been given a good deal more press than cooperation. But certain superficially weak organisms have survived in the long run by being part of collectives, while the so-called strong ones, never learning the trick of cooperation, have been dumped onto the scrap heap of evolutionary extinction.

Though initially skeptical, most biologists now agree with Margulis that evolution took a sudden jump when microbes coupled symbiotically in response to the "holocaust" resulting from the worldwide release by cyanobacteria of a waste product toxic to most bacterial life, including to the cyanobacteria themselves. The polluting toxin was oxygen. The "oxygen holocaust," as it is called, caused massive bacterial death and forced mutations that created new strains. Some bacteria went underground, out of the way of the deadly gas; others developed the ability to "breathe" the oxygen; others engaged in feedback relationships that led to a brand-new step in evolution.

Margulis speculates that the stage was set for symbiosis when one of the cyanobacteria that were creating the oxygen holocaust entered another bacterium in search of food. The host cell moved to protect itself from the sudden presence of oxygen in the cell by forming a nuclear membrane around its DNA, and this created the first nucleated cell.

A second invasion—this time by rod-shaped oxygen-*breathing* bacteria into a host—set off a distinctly symbiotic change. Margulis theorizes that in fighting off the invasion of the oxygen breathers, the host ended up forming feedback links with the invader and the invader stayed, turning the feedback into an arrangement with great benefits. The relationship bestowed on the host the ability to use oxygen as an energy source and, in return, gave the rod-shaped invader a permanent supportive environment. Symbiosis is testimony to the principle that an autopoietic structure will change in order to remain the same. It also demonstrates one of the curious ways that feedback coupling takes place: Here, the attempt to reject an intruder created an interaction that created a marriage.

According to Margulis's theory, the symbiotic mating between the two bacterial strains eventually became so complete that only telltale signs remain of the intruder's separate origin. One of these is the fact that the present-day descendants of the rod-shaped, free-living interlopers—called mitochondria—are a permanent part of our cells, yet they continue to possess their own separate DNA.

Margulis believes that the plant kingdom was born in a similar process when nucleated host cells were invaded by the sun-loving, oxygen-producing cyanobacteria. The resulting feedback interaction "convinced" the cyanobacteria to stay on as chloroplasts and left the new cell with the ability to make energy from water and sunlight, and then, along with the cell's mitochondria, to breathe this former toxic waste. Chloroplasts also have their own DNA.

According to Margulis, fast-moving, corkscrew-shaped spirochete bacteria constituted another intrusion-turned-marriage. If she is right (many biologists don't accept this part of her theory), spirochetes entered into a particularly varied feedback relationship with their host cells. They became flagella and cilia, giving the new nucleated cells mobility. They also became microtubules, strandlike structures inside the cell that carry out a number of functions, from transporting chemical messages and secretions throughout the cell to orchestrating the division of chromosomes in the nucleus. Margulis thinks that in the course of evolution, microtubules in cells also evolved to form axons and dendrites—the business ends of neurons. Thus the early feedback between spirochete and host cells may have led eventually to the development of the brain. It was an ironic fate. The spirochetes are known for their rapid mobility. In one respect the process which transmuted them into brain cells forced them to sacrifice this identity and stay put. On the other hand, their former identity is retained. For packed together and essentially immobile in our skulls, they became instruments of the most rapid-transit feedback network in the history of the planet. Now, in a flicker of electrical motion, they spin not through primeval mud but through the furthest reaches of space and time—as the lightning-fast mobility of human thought.

The symbiotic feedback arrangements that gave cells the ability to move, conduct photosynthesis, and use oxygen to chemically "chew" their food eventually led to still other types of feedback arrangements—for example, sex. Margulis and Sagan say, "Sex, like symbiosis, is one expression of a universal phenomenon, the mix-match principle. Two well-developed and adapted organisms or systems or objects combine, react, redevelop, redefine, readapt—and something new emerges."

Eventually the new symbiotically evolved, sexually reproducing cells coupled together and began to specialize in creating new functions. A nucleated cell with cilia may have joined a second cell, thereby freeing that cell's microtubules to develop in other ways—into a sensing apparatus, for example. The long evolution of the multicelled plants and animals had begun.

Margulis concludes that though we may think of ourselves as autonomous beings, we are, from our toes to our brain, a collection of microbes bound together by symbiotic cooperation. In fact, all life is a form of cooperation, an expression of feedback arising out of the flux of chaos. According to this approach, the Yellow Emperor's kingdom was built and is

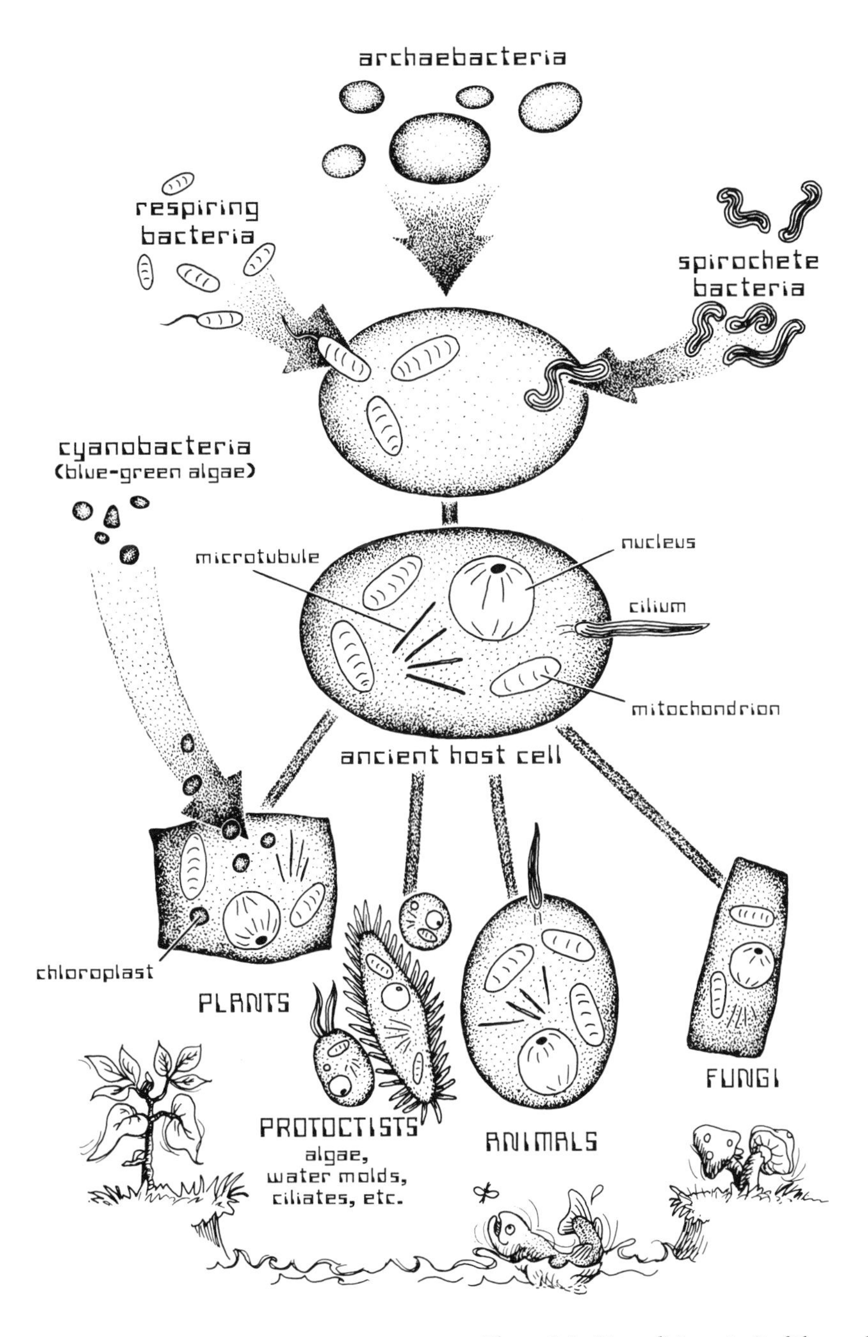

Figure 2.1. **Margulis's portrait of the symbiotic evolution from bacteria to multicellular organisms. We evolved as a collection of cooperating microbes, she claims.**

preserved not out of deadly combat but out of ever-expanding harmony.

By showing convincingly how cooperation is a powerful mechanism for evolutionary change, Margulis adds her voice to the rising tide of theorists calling for a new view of evolution. Though Darwin's original theory can certainly be interpreted to accommodate a picture of cooperation among organisms, the popular and scientific approach to evolution has long emphasized precisely the opposite —that the balance of nature is the result of intense competition among organisms leading to the "survival of the fittest."

But a relatively simple shift in emphasis can lead to a dramatic shift in worldview. Philosopher Robert Augros and physicist George Stanciu have attempted to bring this shift into view in a series of arguments and illustrations presented in their book, *The New Biology*.

Illustration: The competitive model of nature predicts that two species of similar animals must struggle against each other for available food and space. But observation suggests that such struggles are in fact extremely rare. For example, two species of cormorants in Britain have found ways to vary their diets and nesting sites so that they don't compete at all. Though both species nest in similar ways, one fashions its abodes high on the cliffs or on broad ledges while the other chooses narrow ledges and sites its nests lower down. Rather than compete, the two species have interacted with the whole environment and each other in such a way that they carve out different niches.

Another illustration: On the face of it, dominance within species makes a strong case for nature's competitive spirit. Wolves, bulls, birds—all reportedly have dominance hierarchies. However, this structure can be looked at from the opposite pole, as an ingenious means to avoid harmful competition and conflict. Dominance matches between males of a species are usually settled as soon as it is established which animal is stronger. Once that is clear, the match breaks off. This arrangement can be viewed as not essentially competitive but as a cooperative way of avoiding conflict in which the stronger individual would win out over the weaker one, anyway, with probable injury to both.

A third illustration: The theory of Darwinian competition among species rests on assumptions that a species' population will grow without limit unless kept in check by nature's ruthless predation and starvation. Darwin himself used "theoretical calculations" to buttress this assumption, which he based on examples of domestic animal populations that had "run wild" (like the rabbit and gypsy moth populations discussed in Chapter 3). But where species occur naturally, they appear woven into their environment in such a way that they regulate their own population numbers. Populations naturally exist in limit cycles. Studies of white-tailed deer, elk, bison, moose, bighorn sheep, Dall's ibex, hippopotamus, lion, grizzly bear, harp seal, sperm whale, and many other species show that populations accomplish this self-regulation by lowering or raising the birth rate or age of first reproduction depending on population density. When scientists have attempted to remove a species from a territory, the population nevertheless remains stable as animals from neighboring territories fill in the gaps (in the next section we see how neurons in the brain do something similar). Thus, it is not so much "nature red in tooth and claw" that keeps populations within bounds as it is that there's an apparent natural size to a population, just as there's a natural size to an individual organism. The population's size depends on the way it is related by feedback to the whole environment of other species and ecological resources. This makes sense because species evolve in the first place through feedback with the whole evolving environment. Unless human beings interfere, it is largely regulatory and nonviolent feedback that maintains a population's numbers.*

Though most scientists would not agree with the shift in emphasis represented by the holistic feedback approach, there is growing attempt to pose a scientific alternative to orthodox Darwinism.

Augros and Stanciu's new biology is one attempt. The noted evolutionary biologist Stephen Jay Gould has made another by reclaiming some of the ideas of the Russian intellectual Petr Kropotkin. Gould points out that Kropotkin had interpreted *Origin of Species* very differently from the way it was interpreted by European and American scientists. Indeed, Kropotkin found in Darwin evidence for cooperation in nature rather than competition, a thesis the Russian outlined in his book *Mutual Aid*. "If we . . . ask Nature," Kropotkin wrote, " 'Who are the fittest: those who are continually at war with each other, or those who support one another?' We see at once that those animals which acquire habits of mutual aid are undoubtedly the fittest. They have more chances to survive, and they attain, in their respective classes, the highest development of intelligence and bodily organization." Gould notes that Kropotkin developed his interpretation of Darwin after trips to Siberia and Northern Manchuria where he failed to observe a bitter struggle for existence between some species of animals. Gould says: "One might argue that the gladiatorial examples [of animal behavior] have been misrepresented as predominant. Perhaps cooperation and mutual aid are more common results of struggle for existence. Perhaps communion rather than combat leads to greater reproductive success in most circumstances."

Even the reductionist Heinz Pagels, the scientist so bitter in his denunciation of Prigogine, came to the view that the Darwinian theory contains a limited and possibly very flawed explanation for the order we observe in biology.

In his book *Dreams of Reason*, published just before his death in 1988, Pagels wrote that since Darwin "we have come to view natural selection, sifting out rare and useful mutations from myriads of useless ones, as the sole source of order in biological systems. But is this view correct?" Pagels cited computer models of genomic (gene) systems designed by Stuart Kauffman of the University of Pennsylvania that suggest, in Kauffman's words, that complex systems of genes interacting with each other "exhibit far more *spontaneous* order than we have supposed, an order evolutionary theory has ignored" (emphasis added). Kauffman thinks this new form of ordering principle in evolution creates a challenge for scientists to "try to understand how such self-ordering interacts with, enables, guides, and constrains natural selection. . . . Biologists are fully aware of natural selection, but have never asked how selection interacts with the collective self-ordered properties of complex systems. We are entering virgin territory."

* Darwin and Wallace both arrived at the principle that competition leads to survival of the fittest after reading Thomas Malthus's essay on population. Malthus saw competition for subsistence as nature's way of separating the weak and slothful from the industrious and productive among humans. Bertrand Russell pointed out that Darwin anthropomorphized nature by using Malthus's laissez-faire economic theory to picture all life as engaged in ruthless competition. Could it be, however, that this extension of Malthus has inadvertently led to a serious distortion in our understanding of nature, including, among other things, the nature of our own violence? For example, is it possible that the human drive to exterminate other species, fight to the death over territory or sex, and make war on our own kind is much less the result of "natural" animal instincts than of the unnatural conditionings of human culture? If we put aside Darwin's (circular) nature-is-like-man-therefore-man-is-like-nature analogy, we may be free to observe that the aggressive and seemingly violent acts of animal predation are not essentially mirrors of human violence. A lion operating in the economy of hunger dispatches a wildebeest quickly, and there are natural mechanisms that put the lion's prey into a state of shock, minimizing suffering. In contrast, the kind of human violence we worry about is seldom directed at eating what it kills and is drenched in suffering. Unlike animal aggression, human aggression is almost always based on ego, an invention of human consciousness and culture. Whatever its faults, the new biology has certainly made it possible to question some previously unquestionable assumptions that have accompanied Darwinian theory.

Focusing on the way living things self-organize and evolve through mutual dependence shifts the emphasis from the traditional concepts of evolution to a new concept which has been called "coevolution." Examples of coevolution are everywhere. For instance, the ancestral corn plant, teosinte, started out as a common self-sufficient grass on the Mexican plateau. Humans selected it and grew it for larger and larger kernels. Now it's no longer self-sufficient but requires human fingers to remove its thick husks. Nor could humans do very well without corn, a major staple. In a dance of symbiotic feedback, the two species coevolved.

Could coevolution replace Darwinian evolution as the primary explanation for how nature changes? Again most orthodox biologists would resist such an idea. Part of their reasoning would have to do with the belief that our increasing knowledge about that alleged building block of life, the DNA molecule, supports a picture of mutations weeded out by competition and passed along by the genes. But even here the coevolutionary feedback approach is mounting a challenge.

Most evolutionary scientists are convinced that within individuals DNA is a deterministic blueprint. How the genetic code expresses itself is, of course, dependent on loops of feedback between the developing organism and the environment, but the limits are assumed to be set by the code. For instance, researchers assert that some people are destined by their genes to be obese. No matter how much they diet, their genetic predisposition will defeat them.

What does it mean to say that genes are deterministic? Gail Fleischaker, a philosopher of science working in collaboration with Margulis, points out that while it is common for biologists to claim that genes are the ordering agent of the organism, that claim is "altogether unwarranted." Changes in organisms can be *correlated* with genetic changes, Fleischaker says, but that only shows that the genetic changes can affect or disrupt the operation of the organism's total system. It doesn't show that genetic structures *cause* the system's operation or run it. No molecule or type of molecule can be said to *determine* the order of the system.

If genes aren't deterministic, then they can't be the key to evolutionary change. This is suggested by a curious fact. For decades researchers have been bombarding the DNA of fruit flies with x-rays and other treatments to cause mutations, producing all kinds of monstrosities and variations. But none of these mutations has been sufficient to create a new species of fly. Incidentally, the neo-Darwinian view that the gradual accumulation of mutations and genetic variation eventually leads to new species hasn't been supported by the evidence, either. As eminent biologists such as Gould have pointed out, the skeletons in rocks tell a tale of new species appearing with relative suddenness, apparently not as a result of accumulated variations. Augros and Stanciu argue that variation (mutation in a species' DNA) is "not the source of evolutionary change that Darwin thought it was. Its function is to allow species to adjust without extinction or evolution." That means accumulated mutations in a relatively fixed genetic blueprint don't in themselves lead to new species. New species arise from some different process.

Another kind of evidence also questions the reductionist image of a genetic blueprint, suggesting the analogy may be false. Contractors constructing a convention center can follow a blueprint that shows them how to organize their building materials. Once the building is up, the blueprint can be used to locate wiring or plumbing or structural supports. But what if the blueprint keeps changing as a result of the day-to-day weather outside?

Something like that is in fact going on in the DNA in our own bodies, according to geneticist Barbara McClintock. McClintock's biographer, mathematical biologist Evelyn Fox Keller, thinks McClintock's discoveries may form the basis for a "revolution in biological thought."

Working with maize (Indian corn), McClintock observed that the genes on chromosomes actually move around or "transpose" themselves; they even appear to change in relation to environmental stress. McClintock proposed the seemingly bizarre idea that the genetic program isn't necessarily fixed in each individual. In the late 1970s other geneticists found what have been dubbed "jumping genes" and confirmed McClintock's earlier research. However, Keller says that at this point most geneticists don't see these transposing genes as implying a revolution—though some are beginning to realize that there's a fundamental contradiction between "the dynamic properties of the chromosome now emerging and the earlier static [reductionistic] view." Keller reports:

> ... no one can yet quite see how to resolve this contradiction. Does it require rethinking the internal relations of the genome, exploring ways in which internal feedback can generate programmatic change? Or does it require rethinking the relation between the genome and its environment, exploring ways in which the DNA can respond to environmental influences? Or does it require both?
>
> Without question, the genetic apparatus is the guarantor of the basic stability of genetic information. But equally without question, it is a more complex system, with more complex forms of feedback than had been previously thought. Perhaps the future will show that its internal complexity is such as to enable it not only to program the life cycle of the organism, with fidelity to past and future generations, but also to reprogram itself when exposed to sufficient environmental stress—thereby effecting a kind of "learning" from the organism's experience. Such a picture would be radical indeed.

This radical picture is apparently developing. In 1988 John Cairns and his colleagues at the Harvard School of Public Health showed that when bacteria lacking an enzyme for metabolizing lactose were grown in a lactose medium, some of them underwent a mutation that subsequently enabled them to produce the enzyme. This mutation violated the longheld central dogma of molecular biology which asserts that information in the cell flows only one way—from the genes to RNA to protein and enzyme. Here the information was going in reverse. An enzyme coded for by a particular gene was *feeding back to change that gene itself.*

So, on many levels, the DNA code seems less a blueprint than an exquisite feedback relay center balancing the negative feedback ability to maintain stability with the positive feedback ability to amplify change. An inhabitant of the edge between order and chaos, DNA feedback is coupled with other feedback inside and outside the individual organism—an instance of the cooperative, coevolutionary process that sustains and transforms the life on the planet.

To James Lovelock, a British scientist and sometime collaborator of Lynn Margulis, the planet itself is a life-form created by all this interlinking feedback. Lovelock has taken the notion of feedback and coevolution to dizzying heights. According to his Gaia hypothesis, the approximately four billion species on earth are coevolutionarily coordinated in such a way that our planet itself is, in effect, an autopoietic structure, what Lewis Thomas calls a giant "single cell."

Lovelock is a freelance atmospheric scientist and the inventor of the electron-capture device used to gather the data on which Rachel Carson based her environmental blockbuster *Silent Spring*.

Back in the 1970s Lovelock was asked by NASA to devise a way of detecting life on Mars. The British scientist proposed looking for evidence of Martian biology in the planet's atmospheric composition. But first, he needed to study a planet where he knew life left traces —Earth. This study led him to some remarkable realizations.

For one thing, Lovelock was struck by unusual composition of the gases that constitute our atmosphere. An example is the simultaneous presence of methane and oxygen. Under normal circumstances, these two gases react readily to produce carbon dioxide and water. Lovelock calculated that to sustain the amount of methane regularly present in our atmosphere, 1,000 million tons of the gas must pour into the air annually. At least twice that much oxygen must be replaced to compensate for the methane-oxidizing reaction. Looking further, he found carbon dioxide was ten times what it would be if the atmospheric gases were allowed to go to equilibrium. Sulfur, ammonia, methyl chloride are all present in huge amounts above equilibrium. The same is true for the percentage of salt in the sea. Millions of tons of it are washed into the world's oceans every year, yet the salt concentration remains stable. The British chemist concluded that the planet's "persistent state of disequilibrium" was "clear proof of life's activity." He found that, in contrast, the Martian atmosphere is in an equilibrium state. He therefore predicted, correctly, that no traces of life would be found by our Viking probes.

Having once conceived of the connection between life and the disequilibrium of the earth's atmosphere, Lovelock continued his studies and learned another odd fact. In the four billion years since life appeared on earth, the sun's temperature has increased by at least 30 percent, indicating a mean temperature below freezing on the early earth. Yet the fossil record shows no such adverse conditions existed. To Lovelock the explanation for this fact, and the other facts noted above, now became clear: The earth's atmosphere from the very beginning must have been manipulated or regulated by life on a day-to-day basis.

Lovelock postulates that the instruments for this regulation are manifold and have coevolved over time. He reported on one of the planet's negative feedback regulators in an article for the scientific journal *Nature*.

Ocean plankton emit a sulfurous gas into the atmosphere. A chemical reaction transforms the gas into aerosol particles around which water vapor condenses, setting the stage for cloud formation. The clouds then reflect back into space sunlight that would otherwise have reached the earth's surface. If things get too cool, however, the number of plankton is cut back by the chill, not as many clouds form, and temperatures rise. The plankton operate like a thermostat to keep the earth's temperature within a certain range.

Lovelock believes that innumerable biomechanisms of this sort are responsible for the "homeostasis" or steady state of the planet. Just as our DNA, temperature, hormone level, metabolism, and the many functions of our own bodies are balanced by an interlocking series of positive and negative feedback loops, so is life on earth feedback-balanced. The planetary organism, the single cell of earth, remains viable by constantly transforming the elements of its own inner structure.

But the planet Earth is not only a homeostatic organism, it is also a potentially evolving one. The earth's atmosphere hasn't just remained steadily suitable for life, it has also changed in ways that have permitted the continual evolution of new forms of life.

Figure 2.2. The two equlibrium fates of the Earth's atmosphere are depicted in lines A and C. A is an atmosphere like that of Venus: The sun's heat is trapped and the planet is hot, steamy, and intolerable for life. C would have left the Earth's atmosphere similar to that of Mars. Chemicals like oxygen would have reacted with each other and become bound. Heat would not have been held by atmospheric gases, and the surface would have turned cold. Even if we supposed that the climate took a middle course, B, life would have been snuffed out due to the cold conditions from our weaker ancient sun. The actual state of affairs was that life managed to create the temperature required for its own survival. The temperature on the drawing is in degrees centigrade.

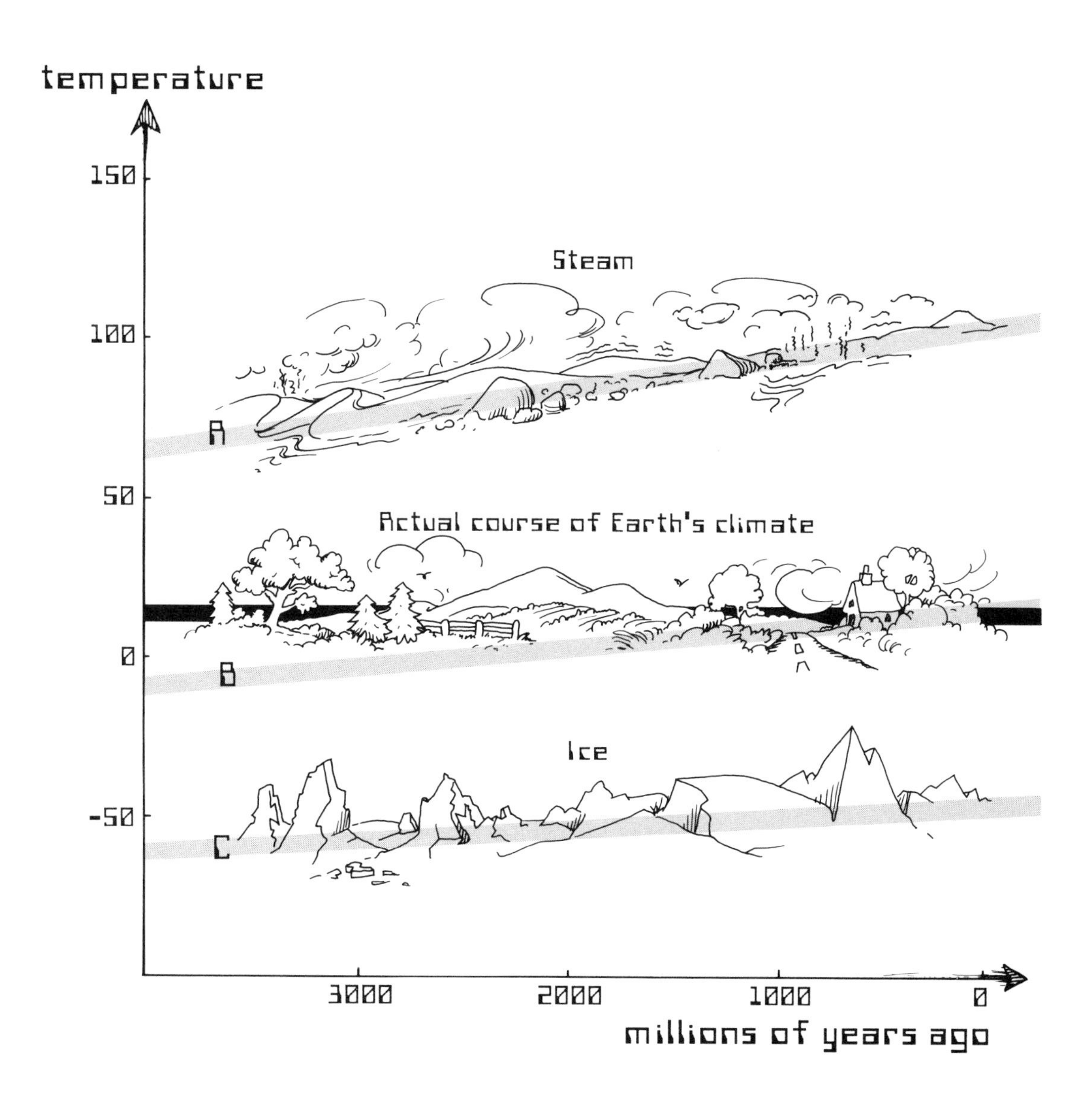

In a soliton wave, the nonlinear correlations of positive and negative feedback loops are exactly counterposed so that the wave remains unchanged as it moves through space. In the earth's feedback couplings, positive feedback sometimes nudges the whole system into a new regime so that evolution takes place. An example of a moment when positive feedback created a new regime was the oxygen pollution crisis caused by the continued activity of cyanobacteria. The toxic accumulation of oxygen in the air could have destroyed all life including the cyanobacteria themselves; instead it fostered evolution. Lovelock says, "When oxygen leaked into the air two aeons ago, the biosphere was like the crew of a stricken submarine, needing all hands to rebuild the systems damaged or destroyed and at the same time threatened by an increasing concentration of poisonous gases in the air. Ingenuity triumphed and the danger was overcome, not in the human way by restoring the old order, but . . . by adapting to change and converting a murderous intruder into a powerful friend." A bifurcation point had been reached and the earth-organism "escaped," in Prigogine's phrase, "to a higher form of order" by evolving an oxygen-using form of life.

Lovelock named his theory of the living, evolving, self-regulating, self-organizing life-form earth "Gaia" after the ancient Greek goddess of earth.

Back in the early 1970s the Gaian idea, Margulis's theory, McClintock's ideas, and the feedback approach in general were received by the scientific community with an attitude close to derision. Though Margulis says she's afraid she will have to die before her theory of symbiosis is totally accepted, it has clearly moved from the fringes to the mainstream.

Lovelock, too, has gained grudging acceptance. His idea that life creates the conditions for its own existence was radical. Until he came along scientists basically believed that life was a mere passenger on the planet, which by chance had the just right environment for evolving biology. But recently his ideas have been taken seriously enough to provide a focus for international conferences and articles in scientific journals. And in 1983 McClintock was awarded the Nobel prize for her research.

Lovelock, Margulis, and McClintock are important figures in a vanguard that is shifting scientific attention away from the traditional theme of "analyzing parts" toward new themes like "cooperation" and "the movement of the whole."

Certainly their approach is not the whole story—there may well be no whole story—but the drastic change of perspective opens up new and exciting insights about the way the universe around us moves.

Systems scientist Erich Jantsch, for example, speculated that the work of Prigogine, Margulis, and Lovelock implies a cosmic scale of coevolution in nature. As we noted, coevolution refers to the kind of interactive influences that occurred between corn and human beings or mitochondria and host. But Jantsch proposed a more all-encompassing coevolution in which the "micro and macro" scales of things, as he called them, evolve together. Bacteria evolve the atmosphere, the atmosphere evolves bacteria. Coevolution couples large-scale and small-scale in a seamless cycle of mutual causality.

Jantsch's notion is unusual in that it runs directly counter to the old scientific belief that nature evolves from the small to the large, from the simple to the complex. Coevolution of micro and macro scales is a fractal idea where both large and small scales emerge as aspects of one totally interconnected system.

Another insight inspired by the feedback approach raises questions about our definition of the individual. It appears that the greater an organism's autonomy, the more feedback loops required both within the system and in its relationship to the environment. This is the autopoietic paradox. The paradox implies that, in a sense, the individual is an illusion. Margulis says, "Really the individual is something abstract, a category, a conception. And nature has a tendency to evolve that which is beyond any narrow category or conception." Could the discovery that individuality is at its roots a cooperative venture be taking us toward a new kind of holism—a holism which will resolve the apparent conflict between individual freedom and collective need?

With their emphasis on universal cooperation as a feature of evolution, it's small wonder

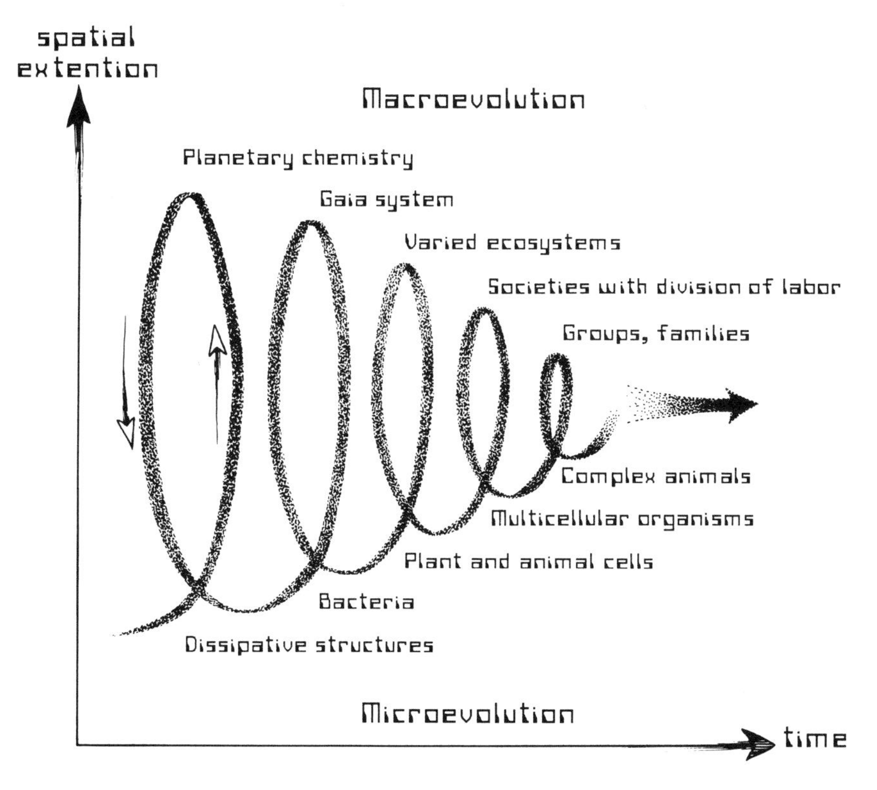

Figure 2.3. **The late Erich Jantsch said in his *The Self-Organizing Universe*, "The history of life on earth expresses the coevolution of self-organizing macro- and microsystems in ever higher degrees of differentiation." Here we see a spiral of coevolution where small-scale changes create large-scale changes and vice versa. Each twist of the spiral leads to greater autonomy on both the individual and collective level. However, greater autonomy also means greater and more complex interdependency. This is the autopoietic paradox. In the drawing above a smooth spiral is depicted flowing between scales, but essentially Jantsch's idea is a fractal one. Coevolution is full of a chaotic order in which large-and small-scale changes mirror each other, jumping back and forth, producing an evolutionary movement that is unpredictable but completely interconnected.**

that Margulis's and Lovelock's theories have been adopted by the New Age movement, environmentalists, the Greens in Europe, and others. But the two scientists have quite different feelings about this popular adulation of their science. Lovelock says enthusiastically, "Gaia may turn out to be the first religion to have a testable scientific theory embedded within it." Margulis complains, "The religious overtones of Gaia make me sick."

The overtones are difficult to avoid, however. The word *religion* comes etymologically from root words meaning "yoking together," and even Margulis can't help implying something almost religious in this sense with her message and with her logic about biological cooperation. For example, she and Sagan make a point of reporting experiments in which microbes that have been enclosed in boxes and placed in the light become more stable as a whole when there are more species and a greater complexity of interaction. One implication is that if complexity among autopoietic structures can lead to stability of the whole, then in saving other species from our greedy meddling, we might in fact be saving ourselves. Margulis openly advocates that if we are to survive the ecological and social crisis we have caused, we may be forced into dramatically new kinds of cooperative ventures. We may perhaps even be pushed toward a unity that has only previously been imagined by religions.

One might speculate that with the advent of the human species, the autopoietic paradox created a new turn in the spiral of the planet's coevolution. The coupled-together bacteria in our bodies and brains have made us independent, autonomous individuals. But at this moment, immersed in a chaotic flux of our own making, we may come to realize that to continue as the individuals we have become we will have to couple on a worldwide scale with each other and with the environment. In their own way, the early bacteria facing the oxygen crisis "realized" the same thing: Cooperate or perish. But this time, if it occurred, the global cooperation would have the added dimension of becoming aware of itself through billions of autonomous human brains. Fittingly, those brains are themselves the sublime creations of feedback and chaos, and day by day harken back to their origins in the first cooperative autocatalytic reactions bubbling at the far-from-equilibrium boundary.

THE NONLINEAR BRAIN

Ilya Prigogine says emphatically, "It's well known that the heart has to be largely regular, or you die. But the brain has to be largely *ir*regular; if not you have epilepsy. This shows that irregularity, chaos, leads to complex systems. It's not at all disorder. On the contrary, I would say chaos is what makes life and intelligence possible. The brain has been selected to become so unstable that the smallest effect can lead to the formation of order." In other words, the brain is the nonlinear product of a nonlinear evolution on a nonlinear planet.

Back in 1987 a *Scientific American* article summed up the current neurophysiological research on memory by reporting that neuroscientists have traced visual memory pathways through six brain areas (sensory area, amygdala, hippocampus, diencephalon, prefrontal cortex, and basal forebrain) with interconnecting feedback loops. This is a large-scale schematic of the kind of nonlinearity that exists at many scales all over the brain. The loops heighten the possibility that bifurcation and the amplification of some input will take place. But is the brain as Progogine argues, really a creature of chaos, a far-from-equilibrium soup simmering on the uneven flame of daily life?

A number of researchers have accumulated experimental evidence that the brain is a nonlinear feedback device, and several neurophysiological theorists are now vying for the honor of providing an overall picture of how cerebral nonlinearity works.

We'll start with the experimentalists. As in other areas of the science of chaos and change, experiments these days include mathematics and models.

Researchers Don Walter and Alan Garfinkel at UCLA devised equations that model the firing patterns of neurons. Linking three neurons together in their model generated evidence of a low-level neural chaos with an implicit order. Walter has said of the model and the brain activity it represents that it is unpredictable in its detail, "but it has *tendencies*."

How this chaotic neural firing is transformed into order is indicated by the research done on actual brains by Walter Freeman and Christine Skarda at the University of California at Berkeley. The two scientists implanted up to sixty-four fine electrodes in the olfactory bulbs of rabbits and monitored the brainwave patterns when the rabbits were given a few molecules of different scents to smell. The researchers found that when an odor was detected, the low-level chaos in this smelling part of the brain self-organized itself momentarily—that is, the firing of the entire bulbful of individual neurons coupled together in a collective way. In fact, the whole system had the appearance of a limit cycle, a different limit cycle pattern for each odor. If the rabbit was introduced to a scent it had never smelled before, the bulb gave out bursts of chaotic activity. If the new odor showed up several times, the bursts gave way to a signature waveform pattern.

Possibly the familiar smell becomes embedded in the fractal pattern of the bulb's low-level chaos, where it is available to be "recalled" through neuron feedback coupling. In these experiments the limit cycle "recognition" of familiar smells was a momentary organized ripple, like a stone thrown into a pond. Here the pond was the ordinary simmering chaos of the rabbit's neuron firing. Achieving the order manifested by this momentary limit cycle is, of course, what the brain is all about. But as Prigogine pointed out, if the brain's order becomes too regular for too long, there is trouble.

Roy King, a neuroscientist at Stanford University, has sketched out the trouble by investigating connections between a neurotransmitter called dopamine and symptoms of schizophrenia such as hallucinations and thought disorders. Drugs that block dopamine were known to lessen these symptoms, but scientists had been unable to find any clear abnormalities in the actual levels of dopamine in patients with schizophrenia.

King and his colleagues at Stanford plugged the known data about dopamine activity into a mathematical model and tried it out on the computer. The model suggests that the key to schizophrenia is the rate at which dopamine is released in the brain. At a certain critical level of dopamine, the neuron firing rate splits into two different rhythms, and the result is a feedback loop gone awry. King describes this brain state as like a fold in Thom's cusp catastrophe. Think of it as a record needle jumping in a scratch. The brain area in question can't get into its normal momentary single limit cycles but keeps skipping catastrophically between two different limit cycles. The schizophrenia victim is suffering from too much order—trapped order—which paradoxically appears, in the epileptic seizure, as a massive attack of chaos.

In the case of epilepsy, a small disturbance in the firing patterns of some brain cells causes a bifurcation. Cells oscillate at one frequency and then are joined by a second frequency; then the first frequency cuts out. This pattern repeats, creating "traveling and rotating waves" that are essentially the same as the scroll-like waves of the Belousov-Zhabotinsky reaction. The conclusion? For the brain, chaos is entirely normal, but the chaos induced by too much order is devastating. One is reminded of Wallace Stevens' line: "A violent order is disorder."

One of the clues to the brain's delicate balance of order in chaos involves a relatively new computer technique that allows scientists to analyze the squiggly plots of patients' electroencephalographic (EEG) scans in much greater detail. Some researchers are using these techniques to look for strange attractors. A. Babloyantz at the Free University of Brussels noticed these complex scans had much in common with fractals and decided to measure fractal dimensions of strange attractors produced by the brain during levels of sleep.

In the waking brain, the chaotic activity of neuron firings is at a low level. But as the brain sinks deeper and deeper into sleep, the chaos becomes more pronounced. However, in the REM levels, when dreaming takes place, the amount of background chaos decreases. Babloyantz believes that the fractal dimension of the brain's strange attractors could provide a measure of the depth of various stages of sleep.

In a similar investigation, scientists at the Center for Nonlinear Studies at Los Alamos have worked out the fractal dimensions for strange attractors associated with different levels of anesthesia. The group also thinks it will be possible to develop a computer analysis of EEG readings to characterize different forms of seizure. Still other scientists want to explore complex brainwaves for the fractal signs of high-level thinking or even creativity.

Could the brain's overall expression, the personality, also be a strange attractor? A psychiatrist at the University of California at San Diego argues that we each possess a unique identity that is written in everything we do. Arnold Mandell claims he has studied individuals' patterns as reflected in firing rate of dopamine receptors, serotonin receptors, and single cells in EEG activity and in the oscillating of behavior—and that he has found a fractal self-similarity among all these indicators.

The frontier of the brain is a vast territory and explorers have only begun to wend their way into the wilderness. Brain models shift in popularity with the frequency of rock stars and it's likely that in a hundred years' time the current maps of the neurophysiological landscape will look as quaint as the sixteenth-century charts of the New World. But a map has to start someplace and among the mapmakers is a growing number of scientists attempting to sketch in the big picture with a nonlinear outline.

One such researcher is Matti Bergström of the Institute of Physiology at the University of Helsinki, Finland. For many years Bergström has worked on what he calls the "bipolar generator" model of the brain. The model divides the brain into an "information" end and a "random" or chaos end, and Bergström says it is the interaction of these ends that produces thought and behavior.

When the retina or other sense organ is stimulated, Bergström argues, the input goes in two directions. One direction is through the cortex, which is organized to convert the stimulus into limit cycle attractors—that is, into an organized form of information.

Input is also circuited through the "random generator." This end is located in the brain stem and limbic system; it takes input from the sense organs and vegetative activities—including the systems controlling digestion and heart rate—and adds them all together. The random generator input is "nonspecific," unstructured—or at least its structure is so highly complex that it contains no information that can be decoded. Bergström says we experience the existence of the random end during those first moments waking up in the morning before we know where or who we are.

Figure 2.4

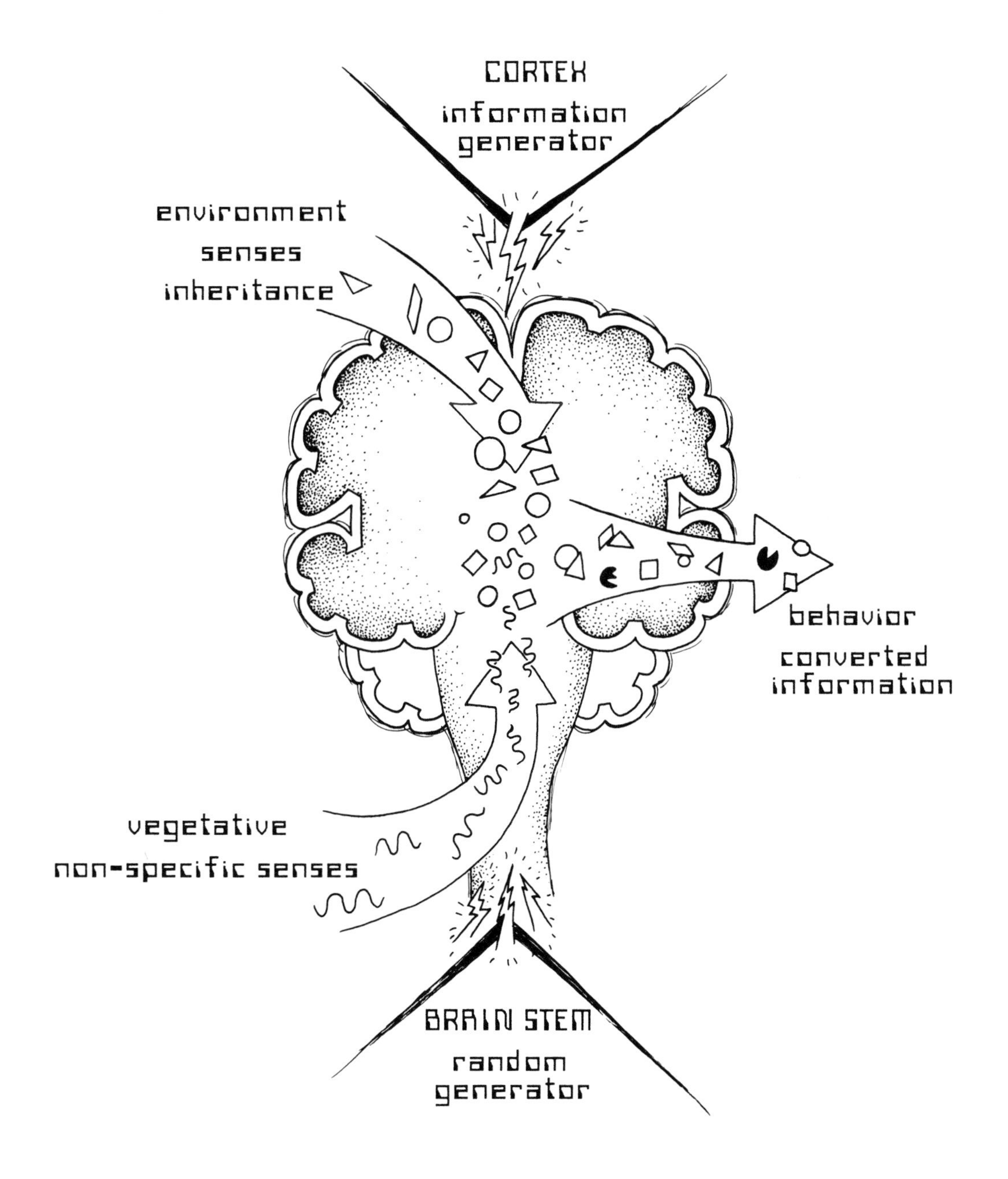
CORTEX
information
generator
environment
senses
inheritance
behavior
converted
information
vegetative
non-specific senses
BRAIN STEM
random
generator

For an instant we have no information, only "being." Our existence and brain activity are nonspecific. Then the information generator kicks in and it all comes back to us.

According to Bergström, when the field of electrical activity from the random generator encounters the patterns produced by the information generator, the result is a "possibility cloud" of limit cycle activity which has been disturbed and reordered by chaotic interference. The possibility cloud therefore contains "mutations" of the information and these mutations engage in a kind of Darwinian fight for survival with the habitual forms of the information. The strongest signals in the total context of signals competing in the brain at that moment will couple together and survive. The output of this contest is a stream of feedback-linked thought and behavior.

Systems scientists William Gray and Paul LaViolette have quite a different slant on depicting a nonlinear brain. They've proposed that thought starts as a highly complex, even chaotic bundle of sensations, nuances, and "feeling tones" which cycle from the limbic system through the cortex. During this feedback cycling, the cortex selects out, or "abstracts," some of these feeling tones. These abstractions are then reinserted back into the loop. The continued abstracting process has the effect of nonlinearly amplifying some nuances into cognitions or emotions, which become organizers for the complex bundles of nuance-filled sensations and feelings.

"Thoughts are stereotypes or simplifications of the feeling tones," says LaViolette. "They're like cartoons of reality." According to this model, the abstracted thought-or-emotions become associated with each other to create larger structures of abstracted thought-or-emotions, which become "organizationally closed." Organizational closure means that the richness of nuance has been summarized (simplified) by thoughts-or-emotions that have a feeling of closure about them. Most of our opinions and knowledge is organizationally closed. We have ceased to pay much attention to the many feeling tones associated with the things we think about or the nuances of our emotional likes or dislikes. But beneath each thought or simple emotion lie layers of sensation and feeling which keep cycling in the brain's feedback loops. Because these nuances keep cycling, the possibility remains that some chaotic or highly charged situation could cause a different nuance to be abstracted and amplified, becoming the organizing thought. Through this process organizationally closed thoughts and emotional responses can sometimes be changed.

Describing how memory is stored and retrieved is a major topic of research and speculation for scientists working on the concept of a nonlinear brain. A number of years ago the well-known neurophysiologist Karl Pribram attempted to answer the problem of how memory is stored by proposing that the brain is a hologram. Experiments and clinical observation had shown that long-term memories remain, even after large amounts of the brain have been destroyed. In one of these experiments, neuroscientist Karl Lashley trained rats to run a maze and then surgically removed different parts of their brains in search of the memory's storage site. He never found it.

Current research reveals that the walnut-sized brain organ called the hippocampus and its associated temporal lobes are connected to memory. Damage to the hippocampus produces profound changes to the memory and impairs the ability to retain memories long term. The hippocampus should not be confused with the *seat* of memory, however, but rather with its retrieval and storage. According to Pribram's theory, the actual site of memory is delocalized across the whole brain.

Pribram proposed that the brain converts sensory input into waveforms. He speculated that these waveforms create interference patterns which can be stored either at nerve cell synapses or in "phase space" all over the brain. This is similar to the way information in a hologram is stored by the interference pattern formed when laser waves are brought together on the holographic plate. In a hologram an image can be retrieved by shining a laser with the same wavelength through the plate. The whole image can also be retrieved when a laser is shined through only part of the plate, though the image in that case is fuzzier. This is analogous, Pribram argued, to the ability of the brain to retrieve information even after large parts of the cortex where the information was stored have been cut out. Pribram has proposed that in the brain a memory is retrieved if a waveform similar to the one that has been holographically stored passes through the brain.

Though experiments have identified some cells in the visual system that respond holographically to spatial frequencies, neuroscientists have not been able to confirm Pribram's holographic waveform mechanism for memory storage. However, while Pribram's theory has not been accepted, as a metaphor, the image of the brain-as-a-hologram may have helped turn neuroscientists toward a more holistic approach to the puzzle of memory. It is also possible that the new nonlinear feedback holism may revive—from a fresh angle—Pribram's phase space idea.

Freeman and Skarda report that in their experiments when the rabbit inhales a familiar smell, the olfactory bulb responds with a limit cycle where "each local region takes an amplitude of oscillation that is determined by the whole. *Each local region transmits the whole* with a degree of resolution determined by its size relative to the size of the bulb" (emphasis added). The limit cycle "memory" for a particular smell may be stored in the whole bulb's low-level chaos or fractal pattern. It is stored there holographically, as it were, because each local region of the bulb contains the whole limit cycle encoded in each of the local region's oscillations.

There is growing agreement that the old theory that memories are stored by individual neurons is incorrect. Rather, memories must arise as relationships within the whole neural network—a sort of phase space of memories.

Michael Merzenich of the University of California at San Francisco has studied monkey brains extensively by planting electrodes. He points out that from monkey to monkey there is considerable individual variation in the locations in the brain where the electrical activity correlated to the movement of the monkey's hand is mapped. In any one monkey these maps of the finger locations also change over time. This means the brain "sites" corresponding to the fingers are not attached to particular neurons but exist as a fluid pattern of relationships. It means that the memory for making a finger movement is not localized at a synapse of a particular neuron but is distributed through a shifting network.

Merzenich found that when a monkey's index finger is damaged or amputated, the areas of electrical activity corresponding to the other fingers drift over to fill in the gap. The drifting of the areas of activity corresponds to the monkey learning to compensate for his disability by using his other fingers. It is to be hoped that the monkey also received compensation for his loss from the experimenters.

If the brain operates by storing its information and functions in networks of relationships among neurons rather than by storing information in particular "wise" neurons or other "hard" structures, then even if some part of the network is destroyed, the rest of the network may retain the information "holographically" in some form.

Scientists working on artificial intelligence have added weight to this idea. A computer

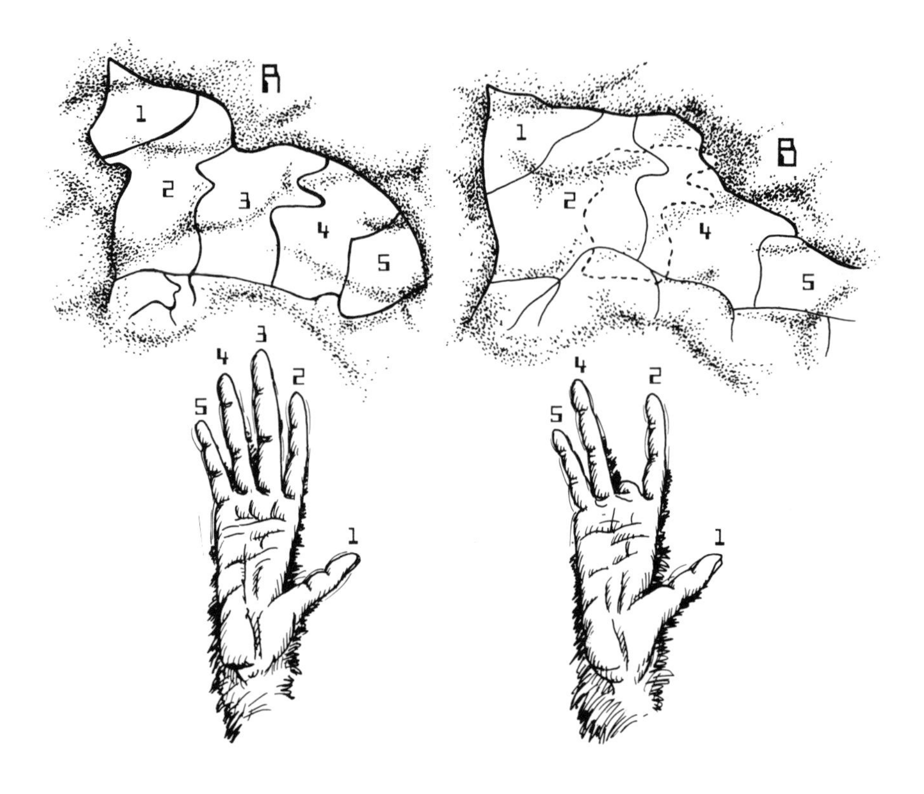

Figure 2.5. The brain area that corresponds to the monkey's missing finger is filled in by the areas corresponding to the other fingers. Scientists are learning that information in the brain is stored in the whole context of relationships among neurons, not fixed to any particular neuron or site.

network called NetTalk has been modeled on neural networks in the brain and has been learning how to pronounce English words. The network consists of 300 computer "neurons" joined together at 1,800 junctions which have volume controls that raise or lower the strength of the signal passing through them. Initially the volume controls are set at random, but after exposure to a list of words and a trial-and-error learning plan, the network self-organizes, becoming better and better at getting the correct pronunciations. Though the network is not provided with any rules for how letters are pronounced in different contexts, it begins to develop and encode such rules implicitly (or holographically) throughout the network. Scientists know the rules are distributed because they can take a "seed" of 10 randomly chosen "neurons" out of the network and reproduce the entire coding scheme. They can also damage the network by ablating or cutting out several "neurons" with the result that while the network's capacity to act becomes a little fuzzy, it still retains the ability to pronounce English words.

The way NetTalk works is obviously very different from the way computers work generally. In a computer which is driven by a program, cut out a few circuits and the whole system crashes. When brain scientists discovered that cutting out parts of the brain does not destroy a memory, they were forced to search for stranger explanations of how the brain encodes its information. Some scientists think that the behavior of computer networks like NetTalk may provide clues to the holographic or holistic organization of neural networks in real brains.

How do neural networks form in real brains? Gerald Edelman, Nobel laureate and researcher at Rockefeller University in New York, has tackled this question with a theory that is steadily gaining acceptance. His theory begins by reaching back into the processes that form the brain in the first place.

It's obvious that there aren't enough genes to govern the location of the brain's 10^{14} synaptic connections. Edelman reasons that the locations of the neurons in an embryo's brain are not preprogrammed by the gene. A few years ago Edelman and his colleagues discovered "adhesion" molecules which guide the randomly growing nerve fibers. Through feedback, these molecules cause the migrating fibers to couple together or self-organize into place, forming columns of small interconnected neuron groups. The exact organization of synapses within each of these neuronal columns and between the columns is unique in every case; no two are wired the same way.

According to Edelman's theory, feedback between the brain and an incoming stimulus "selects" certain clusters of these columnar groups as the brain's response to that stimulus. "Selection" here means that at first many of the neuronal groups respond to the stimulus, but after a while some connections within and between the groups are strengthened by the stimulus while others die away.

To prove his point Edelman's team constructed a computer simulation of a network of neurons randomly connected. Stimulating this network caused some neurons to spontaneously develop positive feedback loops and form clusters of strongly connected cells. In the model, cells that are not stimulated together or that don't have enough connections don't join the clusters.

In a real brain like the monkey's, feedback between the network of neuron groups and the environnment is continual. This suggests that even though information such as that involved in experiencing a sensation or making a movement is embedded in a particular set of neuron relationships today, it may be shifted slightly and embedded in a slightly different set of relationships tomorrow. If Edelman's ideas were applied to memory, they might explain why it's easier to remember where you left your wallet when you can reconstruct the context of your thoughts and movements. A memory, like a sensation, is not an isolated bit; it is a pattern of relationships. Edelman's model might also explain why our memory of a past event transforms over time. The memory floats in an undulating sea of relationships that are continually, if subtly, changing.

The nonlinear approach to the brain has had a major effect on the worldwide effort being made by computer scientists to create in microchip test tubes an "artificial intelligence" (AI).

Psychologist J. Z. Young thinks Edelman's brain model offers the best hope for inventing a "selectionist machine" that would evolve its connections and hierarchy through interaction with the environment rather than through being programmed. Young envisions that such a device "during its prolonged life, might gradually acquire enough experience to generalize about the properties of the world and, as a result . . . show evidence of hopes and beliefs about the future."

There are many different schemes currently being tested in AI research and Edelman's brain model is headed in the same general direction as the popular "connectionist" strategy. Connectionists hold that computer circuits should be wired like neurons with microchip cell junctions (synapses). The computer's programs should not be a logical set of instructions for producing predictable results, according to connectionists; they should instead be merely instructions for varying the strengths of connections between processors, thus encouraging the machine to form nonlinear networks. The connectionist theory is that if all these conditions are fulfilled in the right way, the nonlinear feedback generated in the machine by the problems humans set it will cause the computer to undergo bifurcations and amplifications such that intelligence will self-organize.

The nets constructed to test the connectionist ideas have been relatively simple. Each transistor representing a neuron in the net responds to input from other transistors by switching on or off or by amplifying or diminishing a signal. Which of these actions takes place depends on the "sum" of the input the transistor receives. So far, one computer built with neural nets has displayed associative memory, which is the ability to retrieve a set of scattered facts about a subject even though the starting question is fragmented or partly incorrect. (Remembering that a person you knew in college wore glasses and then remembering other facts about that person is an example of associative memory.) Another example of a computer neural net is NetTalk, with its ability to teach itself to pronounce English.

While powerful digital computers can also perform the tasks neural nets have accomplished so far, neural nets do these tasks more quickly. Neural nets hold promise, but at this point they are only rudimentary forms of the high-level dynamics of the living brain. Freeman and Skarda criticize the connectionists on the basis of their own findings about memories in the olfactory bulb. There, they say, memory depends not only on the interconnectedness of neurons but on a background of chaos. The chaotic pattern to which the olfactory bulb returned after each recognized smell was never the same. The brain's chaos, therefore, makes it quite unlike the precisely weighted connectionist networks. Freeman says that chaos is what "makes the difference in survival between a creature with a brain in the real world and a robot that cannot function outside a controlled environment." Connectionism moves away from the digital logic of computers, but, Freeman and Skarda seem to ask, does it move far enough?

So it remains to be seen whether the connectionist route to AI can succeed. Nevertheless, it is significant that scientists are now pinning their hopes on the nonreductionist aspects of complexity in order to solve the problem of making a machine that can think. Clearly science has come a long way from the days when the predictable, rational aspects of machinery were believed to be the very image of the universe.

NONLINEAR FUTURES

Much of what we've been talking about in this chapter could come under the general heading of a "systems approach to reality." Systems theory is not as gray or mechanical an idea as it sounds. In fact it can be quite lively. One key to systems is nonlinear feedback—and as we've seen, nonlinear feedback can turn the simplest activity into the complex effloresence of a fireworks display.

The systems approach has taken the form of many species of theories that have evolved over the years. There is general systems theory pioneered by the late Ludwig von Bertalanffy; the cybernetic tradition begun by Norbert Wiener, and the servomechanistic or engineering tradition represented by MIT systems theorist Jay Forrester.

In its various forms and hybrids, the systems idea has been infiltrating virtually every discipline. Departments of systems have sprung up in universities all over the world. Futurists like Alvin Toffler, John Naisbitt, Hazel Henderson, and Marilyn Ferguson have proclaimed that the systems outlook is the wave of the future. Nobel prize economist Herbert Simon announced in 1978 that he had abandoned traditional economic theory and was converting to information and systems theory. However, despite the enthusiasm, the systems approach is still a young science that has yet to prove itself as more than a clever new way of looking at things.

Above Peter Senge's desk at MIT's Sloan School is pinned a drawing by his young daughter. It is a swirling spasm of lines, a portrait of chaos, on which she has printed in a preschool hand, "Daddy at work." Chaos and uncertainty are indeed part of the work Senge does at the Systems Dynamics Group. One of a new breed of social scientists, he can serve as our example of the kind of approach systems theorists are taking. Like other systems theorists, he is eager to explain how his brand of the systems view works.

The "system dynamics" idea got its start with Senge's colleague, Jay Forrester, an engineer involved in inventing the core memory for the computer back in the early 1950s. Forrester became interested in applying the engineering concepts of systems to the complexities of social science, and he adopted the new computer as a tool.

Since founding the Systems Dynamics Group, Forrester and colleagues have taught dozens of corporations and municipalities to deal with management problems through nonlinear "modeling."

We all have countless models in our heads about how things work. "If your car starts to skid, turn your wheels in the direction of the skid"—that's a model. "Spare the rod and spoil the child"—another model. Some of our models involve feedback but generally not the kind of iterated (positive) feedback that makes for nonlinearity. In business and economics the theoretical models used for planning have traditionally been linear. "Increase the sales force and we'll increase the number of sales," or "Take the growth rate for the last five years and project it for the next five years after compensating for population declines."

But linear models are notoriously unreliable as predictors, which is their usual function. Forecasts don't work out. The population suddenly starts to grow or moves to another part of the country or starts buying less of a product because of some unforeseen reason, such as a gas crisis. Attempts to make predictions suffer a chaotic fate. The predictions fail because the models can't take in the whole of how the elements in sensitive dynamical systems interact.

System Dynamics' answer to this modeling dilemma was to make the essence of the model nonlinear and to shift the emphasis away from prediction.

Nonlinear models differ from linear ones in a number of ways. Rather than trying to figure out all the chains of causality, the modeler looks for nodes where feedback loops join and tries to capture as many of the important loops as possible in the system's "picture." Rather than shaping the model to make a forecast about future events or to exercise some central control, the nonlinear modeler is content to perturb the model, trying out different variables in order to learn about the system's critical points and its homeostasis (resistance to change). The modeler is not seeking to control the complex system by quantifying it and mastering its causality; (s)he wants to increase her "intuitions" about how the system works so (s)he can interact with it more harmoniously.

Thus, the development of the systems model exemplifies the shift that the science of chaos and change is making from quantitative reductionism to a qualitative holistic appreciation of dynamics.

How is a qualitative model made? When they work with complex organizations such as corporations, System Dynamics modelers try to identify the written and mental concepts the people in an organization are using when they do their work, the organization's rules and policies, the actual behavior of people in the organizational setting, the organizational structure, its purpose, and numerical data such as how many people are working and when they work. The goal is to see what kinds of loops these elements form.

"Initially clients are skeptical," Senge says, " 'You can't model this; this is not just a system of hard variables. We are talking about innovation, passions of man, all sorts of subtle, unmodelable things.' Their first position is almost always cynicism. But after a while they get enthusiastic. They see you *can* model the psychology and the subtler dynamics that go on in an organization. They find that if you can talk about something clearly, you can usually model it, and they get enthused about modeling the subtler dynamics that everybody knows are important."

The tangle of feedback loops is often immensely complex, of course, but the computer

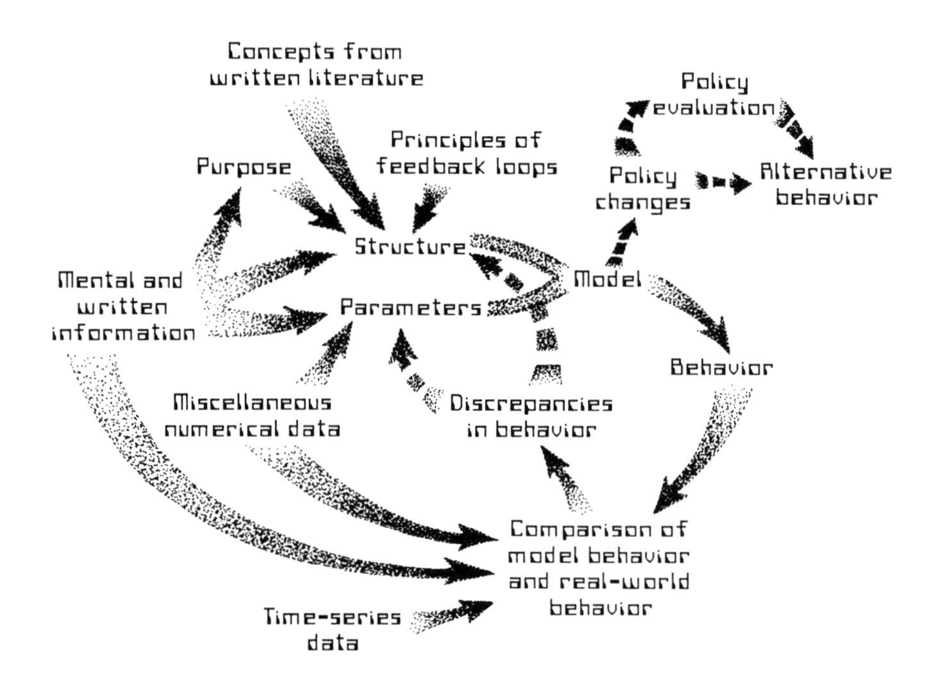

***Figure 2.6.* A picture of the process of making a nonlinear feedback model is itself a nonlinear feedback process.**

can handle that. Nonlinear equations are assigned to the loops to indicate the precipitous things that happen as values are powered up ("loop gains") or diminished.

What is purposely left out of the model are the "historical," or "time-series," data used by linear modelers to compute the ups and downs of past trends the organization has experienced. The nonlinear modeler uses the time-series data not to make the model but to check it. By running the model on the computer, the modeler can see how close his or her picture of the organizational feedback comes to behaving the way the actual organization behaved historically.

One advantage claimed for a good model is that you can change the values in different loops, run the simulation on the computer and see what happens. You can try out a policy change, watch the effect on the system of adding staff or cutting staff; you can experimentally change the relationships of different elements, even gauge the possible result of a difference in employee morale or attitude.

Because it's difficult for a human mind on its own to visualize any more than a very few loops, the computer is indispensable to the modeling process.

By studying systems' complex and varied forms, systems theorists have developed a long list of systems' principles. Below are a few, summarized by Peter Büttner, an executive for the Boise Cascade Lumber Company and a former student of Senge's at MIT:

- To permanently change a system you have to change its structure.
- In any given system there are very few "high-leverage points" where one can intervene to produce significant, lasting changes in the overall behavior of the system.
- The more complex the system, the farther away cause and effect usually are from each other in both space and time.
- It doesn't take very many feedback loops before it gets tough to predict the behavior of a system.
- Neither the high leverage points nor the correct way to move the levers for the desired results tend to be obvious.
- "Worse before better" is often the result of a change of a high-leverage policy in the "right" direction; therefore, any policy change that produces better results immediately should almost always be suspect.

In the past two decades all kinds of large-scale models have been fired up following the Systems Dynamic Group lead. These models of worldwide feedback systems generally have only a few elements and are fairly simple given their scope. Perhaps the best known is the simulation developed in the 1970s by a group of economists, population scientists, and other experts calling themselves The Club of Rome. Directly inspired by Forrester, the group developed a global model, including feedback relationships among the elements of world population, resources, food production, industrial production, and pollution.

The main conclusion drawn from the simulations probably could have been arrived at by common sense: A world economy based on continued growth in all sectors, or even in some sectors, is doomed to fail eventually, probably causing some catastrophic collapse. The model wasn't a prediction of a collapse in a particular time—a fact generally misunderstood. It simply demonstrated graphically that no matter how the variables were manipulated, the growth assumption would eventually lead to a global disaster.

The reason is that all the world's systems are coupled together in feedback loops, and resources are limited. Remember Verhulst's nonlinear addition to the exponential growth equation and the sudden fall it caused in the population of Alice's worms?

One of The Club of Rome modelers, Donella Meadows, notes that the nonlinear coupling of economic factors leads to the inescapable conclusion that "no part of the human race is really separate either from other human beings or from the global ecosystem. We all rise or fall together."

Hazel Henderson believes that the unlimited growth mentality which has dominated world economies is the result of economists' linear approaches to a nonlinear world.

What's the solution to the nonlinear dilemmas that have already begun to affect our standard of living? Many systems theorists are advocating we take a lesson from the mitochondria and spirochetes—learn to cooperate in a new way.

Such cooperation may already be occurring, some systems proponents believe, in the form of an innovative social organism that has flowered across society—networking.

Networking has always existed in some form, as a means for people to communicate with one another outside of the usual hierarchies. But the new networking organism is conscious and entirely feedback-driven. Its sudden evolution appears to come from a growing realization that in our complex world, old societal hierarchies and reductionist control structures aren't working.

Senge says that in most organizations there is a kind of game played in which "subordinates pretend they are being controlled and superiors pretend they are controlling." But the ultimate irrelevance of hierarchy shows up dramatically when an airliner crashes because of a malfunctioning $2.00 bolt. The person who manufactured the bolt was at the bottom of the hierarchy of people who built the plane and yet was sufficient to topple the hierarchy.

The realization of the folly or illusion of hierarchy has helped nourish the expansion of networks, which many social commentators like Naisbitt, Toffler, and Henderson consider the form of the future. In his latest book, *Thriving on Chaos*, management consultant Tom Peters advises managers that in today's volatile world markets the only way to flourish is to "love chaos" by creating a highly nonlinear hierarchical environment within the company. Involve everyone in everything in order to foster creative breakthroughs, Peters preaches. His earlier book, *In Search of Excellence*, popularized the concept of management as networking—"management by walking around." Japanese economic successes also offer dramatic illustrations of successful management that encourages nonhierarchical feedback systems among workers.

Extremely effective global networks have sprung up not tied to any country or social hierarchy. Amnesty International, Greenpeace, and the Coalition of Concerned Scientists are examples. The Green Party in Europe considers itself a nonhierarchical network and abides by the motto "act locally, think globally," in fact the watchword of many networks. Marilyn Ferguson has called networking "the Aquarian conspiracy." Robert Theobald, economist and founder of a network for "social entrepreneurs," says, "linkage and networks are going to be the primary and recognized way to get things done in the future."

William Ellis, originator of TRANET, an appropriate/alternative technology network, is even more visionary: "A future world government can be pictured as a multidimensional network or networks which provide each individual with many optional paths through which he can provide for his own well-being and can participate in controlling world affairs."

Ellis describes TRANET as a structure "composed of links between nodes. It has no center. Each member of the network is autonomous. Unlike a hierarchy no part is dependent on any other. Various members draw together for special projects or on different issues, but there is no bureaucracy demanding action or conformity."

Jeffrey Stamps, coauthor of a networking guide, defines the new networks as "webs of totally free-standing participants." Thus, the cooperative flora evolving and spreading in adaptation to the current world atmosphere appear to have autonomy at the level of the individual "cells" (the network's members).

Like the bacteria that were forced into cooperative ventures by the accumulating atmospheric oxygen, networks appear to be born in a global atmosphere—almost pollution—of information.

Some networks are formed mainly to trade information among people with common interests. Others are expressly designed to create the kind of informational flux that will cause bifurcations and new forms.

Roy Fairfield is an inveterate networker and one of the founders of Union Graduate School, a networking experiment in graduate education begun in the late 1960s. Though fully accredited, the university has no campus or library and offers its far-flung doctoral students instead a "core" graduate faculty skilled in making connections with other students and in keeping the intellectual pot boiling with ideas. Fairfield flies around the country meeting students, and also corresponds via a constant and voluminous stream of letters, haiku, clippings, reading suggestions, and allusions to other students who might have relevant ideas. He says, "I do not make demands in exchange for what is shared." His vision of education is that through networking something creative will happen. He sees networking as a way of maintaining a low-level chaotic substrate so that—as in the brain—the chaos will from time to time give birth to an intellectual self-organizing structure.

Evidently, good networking requires hard work and dedication to the faith that something will come of all the sometimes meandering, nonlinear activity. Networking organisms have died by the hundreds from disuse and they appear to be delicate and transient entities. Perhaps the feedback coupling in these creatures is too weak or loose. Or a fleeting life is the natural fate for a network, allowing its members to move on to other networks. It's also possible we may just not yet have evolved this cooperative species in its most viable form. How can such structures become autopoietic? Undoubtedly, there is still a great deal to learn about nonhierarchical complex order.

Senge, for one, believes that we are only just beginning to understand how to handle such complexity on a social level. He says that when he teaches people how to model systems he starts with "a degree of complexity just within bounds of your conscious ability" and then escalates the complexity until people dimly grasp the whole without actually being aware of it. He thinks learning to handle complexity means learning to live more intuitively, because intuition is the key to making significant changes in complex systems, helping them evolve, and evolving with them.

"At the deepest level of system dynamics we are trying to cultivate a unique intuitive/rational sense of when we are getting close to a critical aspect of a system. You can really *feel* it sometimes, you know when you are getting close to a leverage point. It rarely has any correlation to the symptoms most people focus on, because in a system cause and effect are rarely closely related in time and space."

The point of people immersing themselves in the complexity is, he believes, to liberate their visions. You want to change the system so that it expresses your unique angle on things. But the problem is you can't do that mechanically because your unique angle isn't a reducible item; it's more of a feel, a nuance. So to get at vision, the system has to be approached as a subtle whole. The task, as Senge describes it, is obviously not an easy one for minds trained in reductionism. He says that "there's an incredible tell-me-what-I-can-do-so-I-can-fix-it attitude" that people have about organizations. "We're trying to teach people the systems perspective and part of that is assimilating the ability to grow from acknowledged uncertainty. You're always in an experimental mode. I think it's enormously powerful. It liberates the vision side of things. It also liberates the intellect. In education it lets people operate in a learning mode rather than in a fix-it mode, which makes them a hell of a lot more effective intellectually."

However, he admits that while people get insights from systems dynamics, they often don't stick with the process. "I think in the back of their minds is the thought that despite their insights, somewhere along the line they're going to get this reduction, this model of the system which then they'll be able to change mechanically. After a while they see there's no end to this modeling process, the intuitive process, and they get discouraged. The nature of what we're doing doesn't fit with their assumption of a reductionistic solution."

Perhaps accommodating our minds to subtle holistic complexity is so hard because we have, as Prigogine says, tried to escape time with predictions. It is an axiom in chaos theory that there is no shortcut to learning the fate of a complex system; you have to actually clock it in "real time." The future is told only in the moment-by-moment unraveling of the present. By facing the limitation—in fact the impossibility—of predictions, we may return to real time as the edge between order and chaos, between the known and the unknown, as the depths of the mirror-worlds.

Chapter 1

The Book of the Yellow Emperor *says:*

. . . the root of heaven and earth.
It goes on and on, something which almost exists;
use it, it never runs out.

NONLINEAR PARADOXES IN THE SMALL

It is a strange place. Living in the depths of the mirror-worlds involves living with paradoxes, as Ilya Prigogine and David Bohm illustrate.

One scientist thinks that the root of the universe lies in chaos; for the other at the root lies order—order of "an infinite degree," Bohm calls it, meaning that he sees chaos as really a very subtle form of order. Both scientists agree on the importance of nonlinearity to their new conceptions of reality, but neither agrees about where this nonlinearity leads them. Significantly, the principal site of their disagreement is the quantum—the realm many believe to be the most fundamental level of reality.

Throughout this book we have seen the deep significance of nonlinearity in nature. While linearity may have dominated the physics of the nineteenth century, today linear systems seem almost the exception. Indeed, the first great scientific revolution of the twentieth century—relativity theory—is firmly nonlinear. The strange blur in this ever-sharpening picture of universal nonlinearity is quantum theory. The mathematics of quantum theory are linear. In fact, what is sometimes called the theory's "quantum strangeness" clusters around its linear features.

The essential paradox of quantum linearity lies in what is called the Quantum Measurement Problem. It goes like this: The solutions given by a linear theory, like quantum theory, are all equally good from a mathematical point of view; indeed, there is nothing to stop a scientist from adding solutions together in various ways to form yet more solutions. The solution to any problem in quantum theory must therefore always be given in terms of linear combinations of different solutions—combinations of different outcomes.

However, in any actual quantum experiment there must always be a *definite* outcome:

A geiger counter clicks, a particle leaves a track on a photographic plate; these are all definite and unique events. But how do unique outcomes emerge out of a theory which deals in all possible linear combinations of results? That is quantum strangeness.

The physicist Erwin Schrödinger illustrated this paradox in a particularly graphic way. He imagined an experiment in which the "detector" of a quantum particle's passage is not a geiger counter but a cat inside a box rigged with a cyanide capsule and random triggering device with a 50:50 chance of being activated when a radioactive isotope emits an electron. If the electron hits the trigger in its "on" mode, it will break the cyanide and kill the cat. (Let's add right away that Schrödinger didn't ever plan to try out this experiment with a real cat; it's simply a bizarre illustration of the curious linear property of quantum theory.)

To fully understand the strangeness of quantum theory, we need to see how this device would work in large-scale classical terms. To do that, we'll replace the quantum trigger (the random device and the emitted particle) with a nonquantum equivalent, a roulette wheel and its ball. After the wheel has been spun we know there is a 50:50 chance the ball has fallen into one of the red slots, which has the effect of triggering the cyanide capsule and killing the cat. If the ball falls into a black slot nothing happens and the cat lives. Now, until we open the box we have absolutely no way of knowing if the cat is alive or dead. We can only make a prediction in terms of probability. But there is one thing we do know: The cat must be *either* alive or dead. Common sense tells us that there can be no other possibilities.

Now consider the quantum case in which a disintegrating atom triggers the cyanide capsule. Again we are ignorant of the fate of the cat until we open the box. And we also know that it must either be alive or dead. Or do we?

The problem is that since we are now dealing with a quantum situation we must use the mathematics of quantum theory—a linear mathematics. This mathematics tells us that a live cat and a dead cat are both equally valid solutions of the quantum mechanical equation called Schrödinger's equation. But since this equation is purely linear it is also possible to have valid solutions which contain combinations of both possibilities—a cat which is partly alive and partly dead! In fact Schrödinger's equation predicts all possible linear combinations of live and dead cats. According to the mathematics, all these solutions are valid—and real. Until the box is opened the cat must live in a curious quantum state of suspended animation.

Of course, experience tells us that when we open the box we won't find a host of cats in various combinations of aliveness and deadness. We'll find a unique solution to the experiment—one cat either alive or dead. The multiple solutions of Schrödinger's equation are therefore said to "collapse" into a single unique description—a dead cat or a live one. (How this collapse occurs is another problem which quantum philosophers debate. Is it a result of the consciousness of the human observer, nonlinearities introduced from the world outside, or multiple universes containing live and dead cats?)

Schrödinger's cat paradox clearly illustrates the split between our own nonlinear world of definite outcomes and the curious linear world of quantum theory. With the box sealed quantum theory demands a linear description, combinations of live and dead cats. With the box open we are back to the more familiar world of unique, nonlinear events. But how are these two descriptions to be reconciled? Should nonlinearity be somehow introduced into the quantum world? We've already seen something of Prigogine's answer. Prigogine is attempting to extend the nonlinearity found in the classical-scale reality of high tides and pumping hearts to the invisible quantum scale because he believes that irreversibility and consequently the arrow of time must exist at all levels. For Prigogine nonlinearity represents the universe's creativity. Through nonlinearity he hopes to demonstrate the fecundity of cosmic chaos. Through nonlinearity and irreversibility he wants to entice us into a way of thinking about the universe that will be a "reenchantment of nature."

Bohm, a world-renowned physicist from Birkbeck College, London, has also attempted to bring nonlinearity down to the quantum, but for other reasons. For Bohm, as we'll soon see, the nonlinearity of the quantum is a mathematical clue to what he theorizes is the innate indivisibility and wholeness of nature. Through nonlinearity Bohm hopes to demonstrate the fecundity of the cosmic order, which he believes exists as an infinite complexity of movement. He calls this complexity the "implicate order"—that is, the order of the whole which is implicit in the motion of each "part." Bohm has spent over thirty years devising his implicate order theory and other interlocked theories in an attempt to break out of the reductionism inherent in the linear approach.

In the rest of this chapter we'll explore Bohm's attempt to show how Schrödinger's paradox can be resolved through the addition of nonlinearities and then look at another attempt, called phase locking. First to Bohm.

Bohm's "causal interpretation" is a proposal that introduces nonlinearity into the quantum theory. Bohm realized that it's possible to write down Schrödinger's equation in a new way, essentially by splitting it into two parts.* The first part describes a sort of "classical electron." The second equation describes a bizarre "potential" in which the electron moves, a kind of infinite sensitivity possessed by the electron (or other quantum particle) to its surroundings, Bohm calls this sensitivity the "quantum potential." Since Bohm's equations are a mathematical transformation of Schrödinger's, they will give the same numerical results as conventional quantum theory. Their meaning, however, is very different.

The quantum potential that dictates the way an electron moves is nonlinear and is determined, in an unimaginably complicated way, by all the matter, all the atoms and elementary particles that surround the electron in question. The quantum potential controls the movement of an electron inside an atom, or as it travels within a piece of experimental apparatus.

Because of the extreme sensitivity of the quantum potential, an electron is constantly pushed into bifurcation points along its path, regions where it may be flung in one direction or the other. So complex is this nesting of bifurcations and wildly fluctuating regions that the result is the indeterminism and unpredictability that characterize the movement of an individual quantum such as an electron, "quantum chaos" as it's sometimes called. But as far as Bohm is concerned, the quantum electron's movement is not one of chance and uncertainty; rather, it is totally determined—but by a potential of such endless complexity and subtlety that any attempt at prediction is out of the question.

For Bohm the quantum potential—which every quantum particle possesses—is an infinitely sensitive feedback with the whole. Picture the electron as an airplane controlled by an automatic pilot. The quantum potential associated with the particle is analogous to a radar signal informing the automatic pilot about everything in the plane's environment. The signal doesn't actually power the plane

* In fact Louis de Broglie had earlier discovered how to split up the Schrödinger equation in this way. De Broglie, however, became discouraged by technical difficulties in the approach and soon abandoned what he called his "theory of the double solution."

but can influence its course profoundly through the information carried about weather conditions, other planes in the area, mountain ranges, airport towers. A change in the information will cause a change in the plane's direction.

In the case of the electron, since all the molecules that make up the apparatus surrounding the quantum system are in a constant state of thermal motion, the electron's quantum potential continually fluctuates in an extremely subtle way. Bohm believes that this fluctuation of the whole of the information field gives rise to the probabilistic results of quantum processes—quantum chaos.

What does this mean for Schrödinger's cat? According to the formulation made by Bohm and his colleague Basil Hiley, before the lid is lifted the cat is always in a definite state, either alive or dead—never both, never in some "in between" state of linear combinations of solutions. We can explain this in terms of the airplane analogy.

During the flight, information from the radar beam is taken into account by the automatic pilot and at some stage the constantly changing data cause the plane to head toward a particular airfield. When it arrives (or doesn't arrive) over an airfield, information about alternative places to land is still imprinted on the radar screen but no longer has an active effect on steering the plane. This inactive information is analogous to the other possible solutions to Schrödinger's wave function equation. The wave function collapse is therefore actually an information collapse. The cat is no more half-dead and half-alive than a plane is landing at two airfields at once. As for the wave function solutions that seem to indicate a cat half-dead or half-alive—they don't actually exist, Bohm and Hiley say. Instead, those solutions represent different aspects of the whole information field that is guiding the electron.

So Bohm's quantum potential theory solves the Schrödinger cat paradox. It also has the virtue of making the quantum world totally consistent with the classical domain. It's no longer necessary for the physicist to make a "cut" between the nonlinear large-scale phenomena and quantum linearity, between determinism and indeterminism. Now the same order stretches from the electron to the galaxy.

The nonlinear quantum potential also helps explain what has been called "quantum wholeness."

Experiments have shown that if you correlate two quantum particles and send them flying in opposite directions, whatever you do to one of them will be "felt" by the other, which will react accordingly—even though the two are separated in space. In fact, Bohm thinks, the two particles are coupled together with all other particles by their nonlinear quantum potentials. The coupling even includes the particles in the measuring apparatus. Thus the whole system moves together and what is done to one particle is instantaneously registered by a change in the whole system, thus affecting the other particle.

Bohm's causal interpretation (the quantum potential) is a feature of his theory of the implicate order. He envisions the implicate order as a vast ground of feedback from which quantum processes emerge and in which everything affects everything else. It is a ground made of what he calls "holomovement." For Bohm the universal ground of feedback exists even before there are any "things" to form feedback relationships. In Bohm's implicate order each thing that we identify as a "part" or object actually enfolds the movement of the whole because it is rooted in this infinite nonlinear feedback ground.

Bohm and Hiley admit that the quantum potential approach doesn't predict results different from those obtained by orthodox quantum theory. But they believe the notion does give a mental picture of what goes on at this scale, something orthodox quantum mechanics doesn't do. It also makes quantum events blend with the kind of large-scale nonlinear feedback we saw in sensitive chaotic systems, in bacterial symbiosis, in the Belousov-Zhabotinsky reaction, and in other appearances of order out of chaos.

PHASE LOCKING

There is yet another way in which the linear paradoxes of quantum theory can be resolved—and Schrödinger's cat can end its many-state schizophrenic existence before being let out of the box. A clue to this approach to quantum nonlinearity lies in the basic collectiveness in nature.

Life and nature abound in systems made of linked individuals. We've seen this in the chemical clocks that involve the coordination of millions of individual molecules; in the slime mold aggregating on signal from large numbers of individual amoebae. Systems of linked individuals occur in the way identical cells in a fertilized egg divide and differentiate into separate organs and then work together to maintain the organism; in the ordered atomic structure of a magnet; in the coordination of electrons in a superconductor. What might be called "phase locking" occurs when many individual oscillators shift from a state of collective chaos to beating together or resonating in harmony.

A familiar example of phase locking occurs in our own bodies when we sleep during the night and are awake during the day. In total isolation from any change in light, clocks, or regular meals, our biological chronometers run on a 25-hour cycle. Once in the light again, the biological clock becomes driven by the 24-hour day and is phase locked onto this frequency. But we have only to take a transatlantic trip to knock out this phase locking and experience the disorientation of jet lag as the body attempts to lock into a different 24-hour cycle.

Women living in close groups such as prisons, hospitals, and student residences tend to synchronize their menstrual cycles. Individual spirochete bacteria begin to undulate in the same rhythm when they come together at a food source. Margulis thinks this group rhythm may explain how spirochetes came to form the cilia of primitive animal cells.

Collective oscillations form limit cycles, far more stable and resilient than a collection of individual oscillations. Individual clocks will wander and change, but a phase-locked collective can resist small perturbations.

At Montreal's McGill University Michael Guevara, Leon Glass, and Alvin Shrier took cells from a chick embryo heart and dissociated them in solution where they continued to beat erratically. But after a couple of days the cells came back together. The cells had been able to phase lock their individual outputs to produce a collective oscillation.

The next stage of the McGill experiment involved inserting an electrical probe into the aggregate and administering either a single pulse or a periodic train of pulses. The heart cells were able to phase lock onto the incoming signals and produce a stable pulse themselves. By carefully modifying the frequency, the experimenters were able to use the phase locking to push the cells' rhythms into a region of period doubling and eventual chaos.

The McGill results suggest that our own hearts beat through a system of phase locking of the individual cells. These collectives of cells are driven by the various natural pacemakers—the nerve nodes—which give out periodic signals. The inherent stability of a phase-locked heart is useful when the animal is resting or asleep. For sudden bursts of activity the heart needs to change the basic frequency of its beating, which is where the pacemaker comes in.

If the cells in the heart can phase lock to produce variable though stable rhythms, then what about the complex nonlinear networks of the brain? As we've seen, the nervous system itself consists of an astronomical number of interconnections and has the potential for exploiting a huge range of orders from limit cycles and solitons to the subtle and varied forms of chaos. Regular phase locking occurs at a number of frequencies, as seen in EEG brain rhythms. There are also rhythms that move across the brain and appear to be coordinated with specific types of activity. Is it possible that such global and local rhythms are present in fractal form, being repeated in smaller and smaller regions of the brain?

Phase locking offers a possible explanation of how quantum level systems might come together to create classical-scale systems. Along these lines David Bohm has made an interesting observation about quantum phase locking.

In the first years of this century scientists were faced with the problem of explaining the negative results of the famous Michelson-Morley experiment. Common sense dictated that if you rush toward a beam of light its speed will appear faster than if you rush away from it. But Michelson and Morley's careful test of this idea showed that the speed of light measures the same no matter in what direction the observer or the light source moves.

As it turned out, the Michelson-Morley result required an exceptional explanation—Einstein's special theory of relativity.

But a year or so before Einstein's key paper appeared, another physicist, Hendrick Lorenz, had suggested that the speed of light is not in fact constant as Michelson-Morley indicated; rather, he said, experimental effects conspire to make the actual change in the speed unobservable. Lorenz argued that clocks and rulers are made out of atoms and these atoms are held together by electromagnetic interactions. When any material body moves it must readjust its internal structure. This readjustment makes moving clocks run slower and measuring rods contract. Lorenz contended that together these small adjustments in the measuring apparatus exactly mask the changing velocity of light that the apparatus is trying to measure.

Einstein's explanation of Michelson-Morley was more subtle and far-reaching. He pointed out that time and space aren't absolute things, as Lorenz had supposed, so there was no true meaning to Lorenz's arguments that clocks "really" run slow and measuring rods "really" contract. Instead, the lengths and times of different systems run at different rates relative to each other.

Bohm has combined Lorenz's discarded arguments with Einstein's relativity to come up with a notion he calls "material frames." He suggests that observers—including laboratories or other collective structures—can be thought of as each defining their own local time and space. In one sense Bohm's interpretation is similar to Lorenz's because the time within a material frame is generated out of the phase locking of matter within that frame. But it is different from Lorenz's idea because there is no absolute background space and time against which these clocks and distances can be measured. Instead, time is a measure of the amount of process that takes place, the ticks of the frame's internal clock. When clocks run slow with respect to each other, it is because their material frames are phase locked differently from each other.

Possibly this phase locking of material frames occurs not just for space travelers voyaging across an Einsteinian universe at near the speed of light. It may even take place on individual and cultural levels. It might be manifest by the fact that people and societies seem locked into quite different "senses of time."

Figure 1.1. Phase locking is like an orchestra of individuals all playing the same tune in sync—but with no conductor.

The phase locking of material frames must begin at the quantum level. But how? The answer, we propose, lies in the transformation that occurs when random individual behavior becomes collective behavior. An analogy to the slime mold will help here.

At one level, when there is plenty of food on the forest floor, the mold acts as a collection of individual cells, each one independent of its neighbors and going about its own business. But when food becomes scarce these individuals merge into a collective identity.

They join to become a single corporate being which moves across the floor of the forest. The slime mold clearly shows this transition between individual and collective behavior. We speculate that something analogous may happen at the quantum level between individual and collective phenomena. If so, quantum phase locking could provide a bridge joining classical, nonlinear reality and linear, quantum reality.

Suppose that quantum objects are like slime mold cells in that they also interact together in a collective way. In their individual natures quantum objects may be accurately described by linear combinations of all possible solutions—combinations of alive and dead cats. But as large numbers of quantum objects begin to act collectively, certain stable, definite properties emerge and the collection can no longer be described by a linear combination of different states. Something like this must take place, we would argue, within living systems. Through phase locking, molecules are built up whose properties lie midway between the quantum and the classical. Such molecules have, on the one hand, certain definite properties and on the other are still involved in quantum processes. Some molecules are sensitive to the input of a single quantum particle, for example.

To return to the paradox of Schrödinger's cat for a moment: Clearly the cat is a cooperative, nonlinear system with very definite properties—it cannot be half alive and half dead. On the other hand the disintegrating nucleus which triggers the bottle of cyanide is a linear, quantum object. However, when it couples to Schrödinger's cat, the system as a whole becomes nonlinear and can reside only in definite states.

The mathematician and theoretical physicist Roger Penrose has also looked into what may happen when a large number of quantum objects are coupled together. Penrose chose to work with the most elementary of all quantum entities, spinors, each of which can take on only one of two possible values. He added these objects together according to the rules of quantum theory until he ended up with a large network of spinors. Penrose asked what happens when two such networks are brought into connection with each other. The answer is that they will see each other in spatial terms, as if oriented at a particular angle one to the other.

What is amazing about this result is that Penrose began in a totally abstract way—not working in any space at all but in a pure mathematical domain. Yet out of the interrelationships of the spinors as they locked together to form larger and larger networks he was able to derive the properties of orientation in three-dimensional space. The properties of space, it appears, are not inherent, not given, but emerge in the large scale out of the cooperative interaction of quantum systems.

In a similar way, we suggest, quantum systems may lock together to create not only space but time and other macroscopic structures. It is therefore unnecessary to draw a line between the linear quantum world and the nonlinearities of our large-scale world. For as quantum systems grow in size they will develop nonlinearities and structures.

In some cases the classical-level structure that evolves becomes relatively stable and therefore—as in the case of our solar system—relatively insensitive to individual quantum fluctuations. But other large-scale systems phase lock in a way that leaves them sensitive and close to a chaotic region. In such cases, the classical collective system is responsive to individual quantum fluctuations so that it behaves chaotically, unpredictably, under the influence of a strange attractor.

When scientists make quantum measurements they amplify a single quantum process, resulting in the change of some large-scale variable such as the dial on a meter or the click of a geiger counter. The result is always unpredictable, as in the Schrödinger cat experiment.

Autopoietic structures such as cats and human beings have evolved to exploit the quantum's individual unpredictability. Our eyes, nose, and taste buds are able to respond to just a few quanta of energy. The human nervous system is both classical and quantum in nature, exploiting processes at the quantum scale in order to achieve large-scale ends such as movement or speech. Thus the tension between individual quantum chaos and collective quantum order is able to create and drive increasingly complex scales of structure.

So, the depths of the mirror-worlds are a strange place (or places). In them, Prigogine finds the solution to the Schrödinger cat problem in chaos and the evolving arrow of time. Bohm solves it by finding signs of an infinite, holistic order, and there are other signs which suggest that the solution lies in phase-locking feedback. It's possible all these solutions are wrong, or that they are all correct. At the very least, they all seem reflections of an ancient tension between individual and collective, certainty and uncertainty, chaos and order. More and more we see how that tension is a creative one.

For a long period Ta'aroa dwelt in his shell. It was round like an egg and revolved in space in continuous darkness. . . . But at last Ta'aroa was filliping his shell, as he sat in close confinement, and it cracked and broke open. Then he slipped out and stood upon the shell, and he cried out, "Who is above there? Who is below there?" . . . So he overturned his shell and raised it up to form a dome for the sky and called it Rumia. And he became wearied and after a short period he slipped out of another shell that covered him, which he took for rock and for sand. . . . One cannot enumerate the shells of all the things that this world produces.

POLYNESIAN CREATION MYTH

Prologue

The emperor of the South Sea was called Shu (Brief), the emperor of the North Sea was called Hu (Sudden), and the emperor of the central region was called Hun-tun (Chaos). Shu and Hu from time to time came together for a meeting in the territory of Hun-tun, and Hun-tun treated them very generously. . . .

THAT MONSIEUR POINCARÉ AGAIN

In a nonlinear universe anything may happen. Forms may unravel into chaos or weave themselves into an order. Could the principles of nonlinearity also apply to the creativity of human beings, our ability to make a work of art or a scientific discovery? Fittingly, Henri Poincaré, the scientist who first gave clues to the way nonlinearity and chaos work on the cosmic scale, also provided powerful insights into the way nonlinear chaos operates inside the creative mind. Once again, Poincaré weaves himself into view to tell us that the tension of the old cosmologies still pertains. Indeed, Poincaré showed that in our creative activity, the ancient tension between chaos and order is forever renewed.

Poincaré revealed his insights about creative process in a lecture at the Société de Psychologie in Paris. Here the great physicist described the curious process that led him to solve the problem of fuchsian functions.

Figure P.1

He explained to his audience that for a fortnight he had struggled with this mathematical conundrum but his efforts had seemed in vain until one evening, "contrary to my custom, I drank black coffee and could not sleep." That memorable night "ideas rose in crowds; I felt them collide until pairs interlocked, so to speak, making a stable combination." It was then he saw an order condense out of chaos.

Poincaré confided, however, that the breakthrough insights of his sleepless night were only the first step. These new "interlocked" ideas, when he pursued them, contained yet a new scale of chaos. Out of that confusion sprang yet another perception of order, this one even more dramatic.

> Just at this time, I left Caen, where I was living, to go on a geological excursion under the auspices of the School of Mines. The incidents of the travel made me forget my mathematical work. Having reached Coutances, we entered an omnibus to go some place or other. At the moment when I put my foot on the step, the idea came to me, without anything in my former thoughts seeming to have paved the way for it, that the transformations I had used to define the fuchsian functions were identical with those of non-Euclidean geometry. I did not verify the idea; I would not have had time, as, upon taking my seat in the omnibus, I went on with a conversation already commenced, but I felt a perfect certainty. On my return to Caen, for conscience' sake, I verified the result at my leisure.

Poincaré told his audience that on reflection, his pattern of scientific discovery seemed to be one of initial frustration, confusion, and mental chaos followed by unexpected insight. He recalled another occasion when this pattern occurred. Disgusted with his failure to solve a problem, "I went to spend a few days at the seaside and thought of something else. One morning, walking on the bluff, the idea [solution] came to me, with just the same characteristics of brevity, suddenness and immediate certainty. . . . "

Though Poincaré did not take his insights into creativity much further, they eventually had a profound impact on theories about creativity.

In his landmark *The Act of Creation*, Arthur Koestler theorized that such flashes of order from chaos as Poincaré described must be a function of a process Koestler called "bisociation," that is, the conjunction of two distinct frames of reference. Koestler took as his prototype example of bisociation the story of the ancient Greek scientist Archimedes. Archimedes had been posed the problem of determining the amount of gold in the king's crown but was frustrated because he couldn't figure out how to do it without melting the crown down.

The story goes that one day Archimedes stepped into his bath and shouted "Eureka!" (I found it!) because he realized from looking at the rising bathwater that he could ascertain the crown's volume by putting the crown in water and measuring how much water it displaced. He was able to achieve this ingenious solution, said Koestler, by coupling two quite different frames of reference—the measurement problem and his bath.

In the case of Poincaré's flash of inspiration, Koestler hypothesized that it was the great scientist's change of venue that allowed him to shift the frames of reference on the problem and so arrive at his sudden solution.

Figure P.2

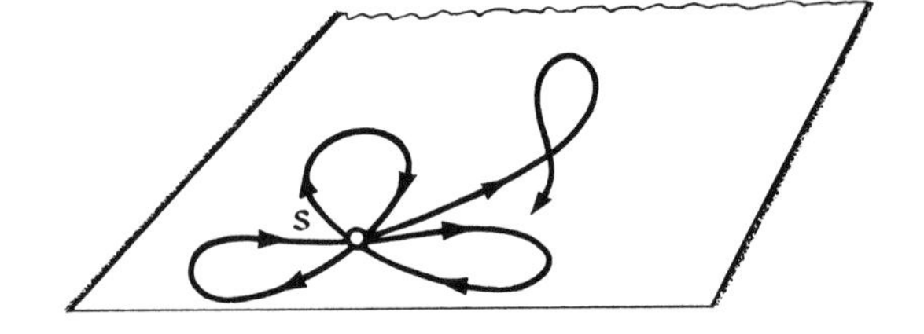

Figure P.3

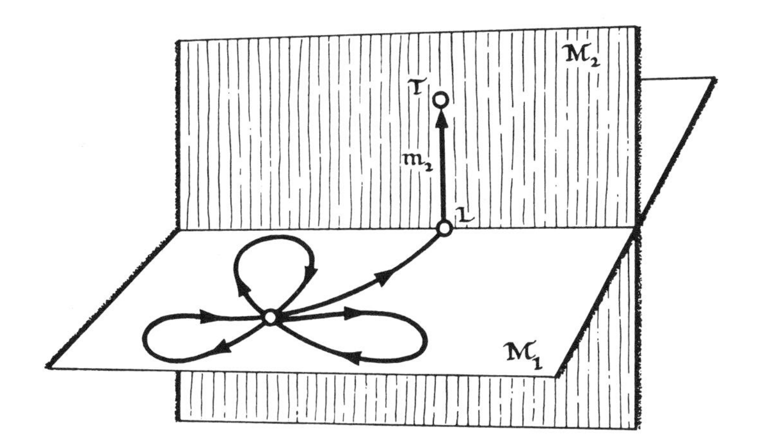

Koestler thought of bisociation as the central process of creativity. His diagrams of the process are a kind of psychological phase space map.

Figure P.2 shows Koestler's picture of the mind wrestling with a problem. The starting point (S) is a kind of point attractor. The intensity of interest pushes the mind away from this attractor in search of the solution or target (T). The initial search involves habitual patterns of thought which act like limit cycles. The mind keeps to these patterns. The target or solution, however, doesn't lie in the same frame (or in this case plane) of reference as the problem; it is not found in the familiar context of previous solutions to related problems.

As Koestler depicts it, the creator's frustration mounts and the search for a solution becomes increasingly erratic, limit cycles breaking down and producing, in effect, a mental far-from-equilibrium flux. At a critical point in this bubbling of thoughts, a bifurcation is reached where a small piece of information or a trivial observation (such as a rise in the level of bathwater) becomes amplified, causing thought to branch to a new plane of reference—a plane that in fact contains the target.

M in Koestler's diagram (*Figure P.3*) stands for "matrix," meaning the context or plane of reference. The L at the bifurcation point is the "link" between the two planes of thought. The link is the factor that is amplified at the bifurcation to create the new order. In Archimedes' bisociation, the link between the two planes may have been seeing the rise in water level in his bath corresponding to the volume of the submerged parts of his body. Poincaré didn't tell us enough about his situation to be able to identify the link which caused him to jump to a new frame of reference when he stepped on the bus. Just physically being away from his workplace seems to have been enough to give him a fresh perspective on all those mathematical elements boiling chaotically in his brain, and the change of scene evidently brought in mathematical ideas from other frames of reference he hadn't included in his initial consideration of the problem. Thus a stray thought about non-Euclidean geometry became amplified and coupled with the fuchsian problem.

A leading creativity researcher, psychologist Howard Gruber of the University of Geneva, has taken Koestler's simple picture of creativity several steps farther. Gruber proposes that creative processes should be thought of not in terms of the coupling of *two* planes of reference but in terms of the coupling of many planes.

As we saw, Poincaré reported that before he came to his final solution of the fuchsian functions he went through at least one previous shift in perspective during his night of coffee-induced sleeplessness. Gruber's research indicates that in the creative process, shifts of perpective go on all the time, at various scales, before the creative solution is finally reached. According to Gruber, many, many small shifts in reference planes couple together, eventually producing a major shift of perception.

The creator's mental effort can be pictured as circling around the problem or creative task, bifurcating to new planes of reference, returning to the old plane, branching to another plane and to planes that lie within planes. This mental effort engenders a far-from-equilibrium flux that destabilizes the limit cycles of habitual thinking. It also couples and phase locks feedback among several planes of reference and begins to spontaneously produce a self-organization.

The ability to jump from reference plane to reference plane while coupling different planes together appears to depend upon the creator's sensitivity to nuance.

NUANCES: AN EXTREME SENSITIVITY

Indeed, a major distinguishing characteristic of a creative person is an extreme sensitivity to certain nuances of feeling, perception and thought. A nuance is a shade of meaning, a complex of feeling, or subtlety of perception for which the mind has no words or mental categories. In the presence of a nuance the creator undergoes what might be called an acute nonlinear reaction. Henry James reported that his story *The Spoils of Poynton* was triggered in his mind when a woman sitting next to him at a dinner party dropped an idle comment about a mother and son fighting over an estate. At that moment James experienced a vivid if amorphous sense of what he called the "whole" of the story that he would soon sit down to write. The woman's words had become amplified in his thoughts by his sensitivity to the peculiar nuance (the complex of unspeakable, uncategorizable feelings and thoughts) that existed for him in the event she was talking about. Every creator is sensitive to different types of nuances. Nuances are like the richness of the boundary area in the Mandelbrot set, the richness of the many scales of a fractal. For a creator, nuances are full of a sense of the "missing information." The pressionist Claude Monet was inexhaustibly sensitive to nuance involving the shifting of sunlight. Virginia Woolf responded strongly to any nuance involving wavelike movements. This nuance led her to some of her greatest novels. A nuance is at first a very private affair. Since its richness isn't described by or contained in the normal forms of thought, it isn't easy to share with other people. To express his or her experience of a nuance, the individual has to create a form which will get the nuance across.

James called any nuance-laden idea or image which provokes a creator into making a new form a "germ." Do scientists react to germlike nuances, too?

Harvard science historian Gerald Holton argues that creative scientists are keenly sensitive to nuances having to do with certain "themata," that is, themes, that they perceive in nature. These themes go deep in the individual scientist's background and often involve a nuance the scientist first felt as a child. Holton connects the discovery of relativity to the rich nuance Einstein felt in the theme of "continuum." Einstein recalled that when he was five years old he was shown a compass by his father. The mysterious power of the electromagnetic continuum in which the compass needle floated inexorably drew him. "Young as I was, the remembrance of this occurrence never left me," Einstein later declared. Holton believes the needle in the magnetic continuum was related in young Einstein's mind to his early religious longings and his perception of an invisible unifying force in the universe. Later on the theme of the continuum in nature seems to have been a nuance-laden germ which ignited Einstein to several scientific projects, including relativity theory and the quest for a universal continuum which he called the "unified field."

The world is, of course, full of potential nuances; it is saturated with shades of meaning, feeling, and perception—experiences for which our languages and logics have no categories. Nuances exist in the fractal spaces *between* our categories of thought. According to the theory of Paul LaViolette and William Gray, nuances circulate all the time from the emotional and perceptual centers of our brains only to become rapidly simplified by our cortex into thoughts that are categorical or "organizationally closed." Everything we regard as our knowledge of the world is organizationally closed. But our wondering, uncertainty, and questioning are full of nuance. In experiencing nuance we enter the borderline between order and chaos, and in nuance lies our sense of the wholeness and inseparability of experience.

One sculptor described her childhood experience of nuance: "A small puddle, iridescent with spilt oil and reflecting a patch of midwestern sky would suddenly expand for an endless split-second to encompass my entire universe."

While most of us pass by such perceptions, or even suppress them because they threaten our customary way of thinking, creators focus on and amplify them. Creators cultivate the ability to live in what Keats called the "doubts and uncertainties" created by a nuance long enough to permit something new to bloom there.

When a germ which contains nuance falls upon mental ground sensitive to it, the result in the creator's mind is a disequilibrium flux of wondering, uncertainty, and wholeness which allows the material being worked with—whether it's scientific data, a landscape and canvas, or a set of characters in a novel—to amplify subtleties, bifurcate to new planes of reference, and form feedback loops among different planes in a process which self-organizes a form to embody the nuance.

Our usual patterns of thought organize themselves around their limit cycles. Asked to make form out of a complex mass of material or to solve a problem, the mind's typical response is to formulate a reductionist or organizationally closed structure rather than letting the material self-evolve out of the fractal dimensions of nuance as a creator would.

What kind of creation is produced by this nuance-grown self-organization? We turn to the world of the creative arts to answer this question.

THE FRACTAL NATURE OF CREATIONS

In *The Monkey Grammarian* Octavio Paz declares: "The vision of poetry is that of the convergence of every point, the end of the road. . . . The dizzying oblique vision that reveals the universe not as a succession . . . but as an assemblage of worlds in rotation."

A poet unfolding nuance is like an equation iterating on the boundary between finite order and infinite chaos. The creator discovers self-similarity. Take as an example of that self-similarity this poem by Pulitzer prize winner Richard Wilbur.

THE WRITER

In her room at the top of the house
Where light breaks, and the windows are tossed with linden,
My daughter is writing a story.

I pause in the stairwell, hearing
From her shut door a commotion of typewriter-keys
Like a chain hauled over a gunwale.

Young as she is, the stuff
Of her life is a great cargo, and some of it heavy:
I wish her a lucky passage.

But now it is she who pauses,
As if to reject my thought and its easy figure.
A stillness greatens, in which

The whole house seems to be thinking,
And then she is at it again with a bunched clamor
Of strokes, and again is silent.

I remember the dazed starling
Which was trapped in that very room, two years ago;
How we stole in, lifted a sash

And retreated, not to affright it;
And how for a helpless hour, through the crack of the door,
We watched the sleek, wild, dark

And iridescent creature
Batter against the brilliance, drop like a glove
To the hard floor, or the desk-top,

And wait then, humped and bloody,
For the wits to try it again; and how our spirits
Rose when, suddenly sure,

It lifted off from a chair-back,
Beating a smooth course for the right window
And clearing the sill of the world.

It is always a matter, my darling,
Of life or death, as I had forgotten. I wish
What I wished you before, but harder.

The poem is constructed as an interlocking series of metaphors, or rather, "reflectaphors." A reflectaphor is any creative device (including, in literature, such devices as irony, metaphor, simile, pun, paradox, synecdoche) that relies for its effect on creating in the mind of its audience an *unresolvable tension* between the similarities and differences of its terms. In other words, a relfectaphor excites a state of intense wondering, doubt, and uncertainty—a sense of nuance. A major reflectaphor in Richard Wilbur's poem is the comparison between two terms: the daughter's effort to write her story and the starling's effort to beat a "course for the right window." These two terms are obviously very different, coming from very different shelves in our mental library. Yet the way Wilbur has juxtaposed them suggests similarities. The tension between the obvious differences and the discovered similarities forces the reader's mind out of its categorical filing system into subtleties and nuance.

The second major reflectaphor in the poem suggests the similarities (and also the differences) between the daughter's effort to write and the father's effort to understand what she is going through as she is trying to write. A third likens the daughter to the house (for instance, by associating the girl's mind struggling over the story with the room at the top of the house where the windows are "tossed with linden"). Here we can see how, though they are different, the terms of a reflectaphor reflect each other, hence the name.

The fourth major reflectaphor is particularly interesting because it entails a metaphor (remember that metaphor is one variety of reflectaphor) which the father rejects in the fourth stanza. There, in his attempt to understand the nuance of his daughter's experiences, the narrator deliberately compares her stuggles to a sea voyage. Just as he says this, however, the girl "pauses / As if to reject my thought and its easy figure." With these words the narrator realizes that the sea voyage metaphor he has used is a cliché. It is a *dead* metaphor, a metaphor which has lost the tension between its terms. Instead of eliciting the depth of his feeling about her struggles, the metaphoric comparison of her life to a sea voyage denies her nuance, categorizes and simplifies her experience, is "organizationally closed." Wilbur, a poet, a writer, is acutely aware that for metaphor to elicit nuance it must be fresh, not dead; it must shock the mind into wonder by opening up a gap between its terms and then bridge the gap with an electricity of nuance. Overuse closes up the gap between the terms of a metaphor because we come to think we "know" what the metaphor means. Comparing life to a sea voyage seems pat because we think we know all about life's rough sailing and "heavy cargo."

Ironically, however, by recognizing the closed and categorical cast of his sea voyage metaphor, the writer who is narrating the poem forces himself and the reader to realize that we really *don't* know what it means to say life is like a sea journey. As a result of questioning the cliché, the sea journey metaphor is injected with wonder and is thus able to make an implicit comeback in the last stanza of the poem, this time brimming over with nuance.

A reflectaphor, with its nuances arising out of an unresolvable tension between its terms, is like a fractal. Fractals, remember, are both order and chaos. They have self-similarity at different scales, but this self-similarity is not self-sameness and is unpredictable and random. The tension between similarities and differences in reflectaphors also creates for us a sense of unpredictability and randomness in the creative work, a sense that what we're experiencing is organic, is both familiar and unknown.

The terms of a reflectaphor are like poles in an electrical device. In the gap between them flows an electricity of nuance. When there are several reflectaphors in the piece, the poles (terms) interact with each other like circuits or feedback loops, each affecting others to create a movement of nuance: It is a self-organizing movement on the boundary of both order and chaos.

This is an abstract map to the poem's wholeness and self-similarity. The feedback shows that everything not only affects everything else in the poem, it *is* everything else. The father is, in some sense, the daughter (he is a writer, for example; he is struggling with how to express nuance at the very moment she is struggling with it); the daughter is the house; the father is the house—and so on. Those items are also *not* each other; their differences are vital, too.

Wilbur's process of creating this poem probably self-organized out of some nuance-laden germ containing his sense that the act of literary creation and parent-child relations are somehow the same. Appropriately, the result of unfolding that germ was a self-similar object. As James said, germs also contain a sense of the whole. That wholeness is embodied in the self-similarity of the finished work, where each part is coupled to, was generated from, and is a reflection of each other part.

Which leads to an additional self-similarity clearly evident in this poem as soon as we realize that in writing it, Richard Wilbur must have been undergoing the same kind of struggle he describes the father and the daughter undergoing. This scale of self-similarity is by no means unusual in creative works. It's conceivable that every great work of art at some scale can be read as a portrait of the mental struggle the artist was going through to create the work itself: *Moby Dick*, "Guernica," the Emperor Concerto seem clear examples.

Through all this self-similar feedback, the artwork reveals that there are worlds within worlds. Yet perhaps the most important scale of self-similarity remains to be considered—

Figure P.4

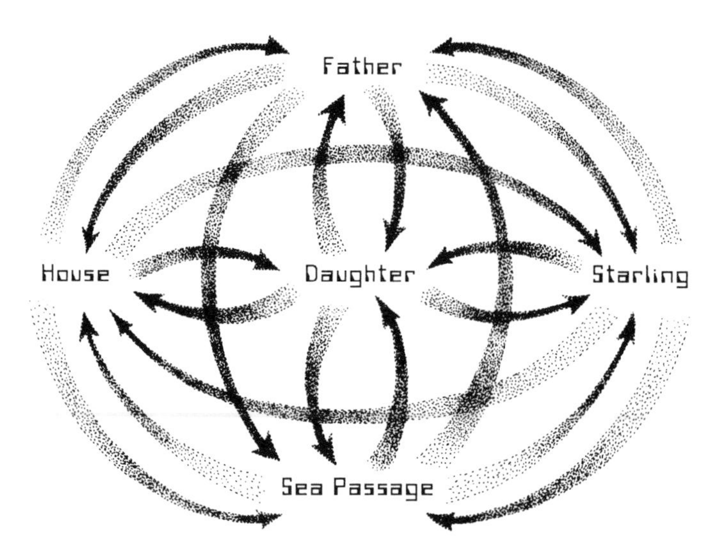

the self-similarity that exists between the artwork and its audience.

In the Wilbur poem this self-similarity shows up as the reader's struggling to connect with the elusive nuances of the poem in much the same way the father struggles to connect with his daughter and the daughter with her story.

A poem such as Wilbur's exhibits reflectaphoric tension at all its scales. The poem exists in the nuances of its metaphors, ironies, paradoxes, images—in other words, in the dimensions between the limit cycle attractors of language. The resulting movement of nuance (the poem itself) is, as we've seen, fractal or (if you like) holographic in that each part reflects every other part—but not exactly.

The kind of fractal/holographic structure of Wilbur's poem is also evident in the Hokusai painting "The Great Wave" which we looked at on page 112. You might want to glance back and note the reflectaphoric tension between the momentary fluid shape of the foreground wave and the solid Mt. Fuji in the background; between the boats which repeat the curve of the waves but are also imperiled by them; between the faces of the boat crews and the flecks of foam.

Hokusai's inclination to make the deep structure of his painting fractal is by no means unusual. Look at this portrait of Maria Portinari by the fifteenth-century Flemish artist Hans Memling.

Notice how the ovoid shape of the woman's eyes is repeated in numerous reflectaphoric variations and tensions throughout the painting: for instance, in her necklace and the peak of her conical hat, even in the curve of her thumbs. In its fractal structure, the Memling painting reveals the paradox of simplicity as complexity and complexity as simplicity.

Once fractal geometry had been discovered by Benoit Mandelbrot, artists began to consciously recognize it as a feature of their art. "With a fractal, you look in and in and in and it always goes on being fractal," says British painter David Hockney. "It's a way towards a greater awareness of unity." Hockney thinks of his own work as holographic and fractal.

Musicians have noticed the connection, too. Composer Charles Dodge, director of the Center for Computer Music at Brooklyn College,

Figure P.5

links fractals to a basic self-similarity that has always existed in classical music. "The awareness of self-similarity abounds in studies of musical structure," Dodge says.

For example, Leonard Bernstein in his Harvard lectures pointed out musical self-similarity from the largest to the smallest scales of musical structure, calling such repeating variations "musical metaphor." And the composer Arnold Schönberg insisted that in a great musical piece "dissonances are only the remote consonances." Nuance, self-similarity, wholeness, and reflectaphoric tension are all implied by Schönberg's statement.

But contemporary composers have not only been able to observe the similarities between fractal geometry and the traditional aesthetic structure of their art, some have employed the actual technology of fractals in some of their compositions.

Pulitzer prize-winning composer Charles Wuorinen said he was inspired in 1977 by reading Mandelbrot's book on fractal geometry. Fascinated by the idea of the "behavior of parts of nature" that it implied, he wrote several pieces using fractal algorithms. One, entitled *Bamboula Squared*, was composed for quadraphonic tape and orchestra and was performed by the New York Philharmonic in 1984. According to Wuorinen, the pieces were generated by finding the "right" algorithm and iterating it as a random fractal. The right algorithm means one that creates nuances by balancing the randomness with self-similar features. The resulting piece forces the listener to constantly interact with the music by recognizing it as a cloud of sounds which are obviously ordered and similar to each other but also are constantly unexpected and different. This perception of the expected-as-unexpected is a vital facet of creative expression. It makes the tension between order and chaos forever new. It is what Paz called "the dizzying oblique vision that reveals the universe not as a succession . . . but as an assemblage of worlds in rotation."

THE ART OF SCIENCE AND OTHER ARTS

David Bohm has proposed that science in the future should move closer to art. He makes two suggestions. First, he argues that instead of science discarding alternative scientific theories in favor of one "accepted" theory, scientists should pursue the possibility that scientific truth, like artistic truth, is a matter of endless nuance, of "worlds in rotation." The root of the word *theory*, Bohm points out, means "to see." Because of reality's infinite nuances there may be many, even opposing, ways to see what nature is doing. Artists, of course, have known this for a long time.

Bohm's second suggestion for making science an art is to have the authors of scientific theories build into them a kind of irony akin to the irony of art. This irony would be a recognition that whatever the theory says about reality is not in fact that reality, because any theory is an abstraction of the whole and therefore is, in a sense, an illusion. Though scientific theories may be quite useful illusions, Bohm believes, the user of a theory should always be starkly aware of the theory's inherent limitations. This, too, is a bow to a reality of infinite nuance. To a certain extent, a good theory should undercut itself, like the Yellow Emperor in his most Taoist mood.

Peter Senge says our future social business and social group activities might best be conducted in an atmosphere of irony and nuance as well. He calls it acknowledging the basic uncertainty. "A reverence for uncertainty," he says, "is one of the unrecognized implications of systems thinking." He links this reverence paradoxically to a visionary quality he has seen in successful corporate executives who have extraordinary energy and ability to convert their personal perception of nuances into a form that has an impact on others.

"I think the reverence for uncertainty is the difference between a creative visionary and a fanatic. A fanatic looks for something that will stamp out the uncertainty. The creative person acknowledges the uncertainty. That person says, 'Here's what I'd really like to see happen. I'm not sure it's possible but I'd really be willing to stick my neck out for this.' "

In Senge's scheme, a reverence for uncertainty leads to a personal vision which is, in turn, linked to the individual's ability to foster the collective action of many individuals. Here we're reminded again of Keats's claim that the capacity to live in "doubts and uncertainty" (in nuance) is the basis of creative power.

The ancients said that the artist's task was to hold a mirror up to nature. What they meant was probably misunderstood by later ages, for the mirror of art has never been a mere slavish imitation of nature's forms and gestures. Rather it was always an Alice-in-Wonderland mirror, as full of play and uncertainty as nature itself . . . a mirror reviving in new forms the ancient tension between order and chaos. Possibly in the bifurcations leading to our future, science and our social institutions will join with the arts in holding up to our turbulent universe such a playful and turbulent mirror. Perhaps in what we've seen throughout this book, that movement is already beginning.

FOREWORD

Humankind is fast approaching a bifurcation point. During this century, reductionist assumptions carried scientists deep into the atom where they liberated the awesome nuclear forces that could spell our doom. However, the pursuit of reductionism into the heart of the atom also liberated important insights into the limits of reductionism. The paradoxes of quantum theory revealed to scientists the mysterious "quantum wholeness," the vast implications of which are only beginning to be explored. But in the meantime most physicists carry out the reductionist program as if nothing had changed. They build larger and more powerful colliders in search of nature's building block parts—quarks, gluons, and the potential primordial force that gave birth to the universe.

In molecular biology the reductionist approach of analyzing reality into constituent parts and reassembling it according to our needs and fancies is now leading to a biotechnological revolution. With recent genetic discoveries it is fast becoming possible for scientists to redesign existing organisms and to generate new ones, introducing the prospect that we will one day transform the planet into a habitat populated with our own creations. Emboldened by our genetic knowledge we may soon be tempted to intervene in even our own evolution.

Nature controlled by human thought is the essence of the reductionist dream. It is a dream that persists, even in the face of its evident failures. The orientation that treats each system as mechanical, made of parts and isolatable from other systems, has given rise to a technology so powerful that it dominates the world. But a direct by-product of that technology is the warping of the planet's environment, including the depletion of stratospheric ozone and the building up of greenhouse gases. Many scientists are now predicting that these aspects of technology and progress will lead to ecological disasters and chaos for our species. But the reductionist dream is not shaken. In a mechanical world, what reductionist science spoils it can also "fix." Thus proposals are being made to slingshot frozen ozone into the atmosphere to repair the damage.

Against this trend rises the young science of chaos, wholeness, and change—a new insistence on the interrelationships of things, an awareness of the essential unpredictableness of nature and of the uncertainties in our scientific descriptions.

Between the holistic and reductionistic points of view, which will we choose? Perhaps a measure of the mounting struggle is the degree to which the reductionist position has appropriated holistic language. It is common now to hear scientists talk of "perspective" reality instead of objective reality, of "creative possibilities" instead of causality, of "likely scenarios" instead of deterministic outcomes, of "useful models" instead of permanent truths. Though such language may seem holistic, that is not necessarily the case, Jeremy Rifkin observes:

> At first glance, terms like "perspective," "scenarios," "models," "creative possibilities" appear to signal a newfound awareness by humanity of its own limitations, of its inability

ever to fully grasp or comprehend the truths of the universe. Not so. It is not humility that animates the new cosmological jargon but bravado. When we take a closer look, the new vocabulary suddenly takes on an entirely new appearance, at once menacing and intoxicating. Perspectives, scenarios, models, creative possibilities. These are the words of authorship, the words of a creator, an architect, a designer. Humanity is abandoning the idea that the universe operates by ironclad truths because it no longer feels the need to be constrained by such fetters. Nature is being made anew, this time by human beings.

Thus the new holistic vocabulary may conceal a traditional reductionist impulse, the impulse of an assembler and manipulator of parts. The co-opting of language indicates that the reductionist urge in science is powerful, so powerful, in fact, that it is almost impossible to think of science without that drive to get to the absolute bottom of things, to find that absolute part, to learn the absolute basis of forms.

Yet the holistic impulse in science is also powerful, a mirror image of the reductionist one. A scientist may pursue the absolute part because he or she wants to see the interrelationships of the whole. The desire to have a reductionist answer is often accompanied by the need to have a mystery to work on. The difference between reductionism and holism is largely a matter of emphasis and attitude. But, in the end, that difference is everything.

In the coming years, the growing struggle between an attitude of unrestrained reductionism and the attitude represented by the turbulent science will be decided. The terms used by proponents of the two approaches will not always distinguish their positions; the issues dividing them will not always be clear—but in time the question will be answered. Will we carry reductionism on toward the ultimate dream (and perhaps the ultimate deception) of turning nature into a mere extension of human thought? Or will we enter the turbulent mirror embracing our limitations and acknowledging our dependencies?

If we do enter the mirror, what will we find? Obviously, no one knows. Scientific ideas of cooperation and inherent unpredictability could usher us into undreamed-of realities and unthought-of activities. It's even possible that these new turbulent realities will be more dramatic than the science fiction futures promised us by the reductionist point of view. Or perhaps the new reality will be manifested mainly by our change of attitude.

Could it be an attitude like the one geneticist Barbara McClintock has taken toward her work? "Basically," McClintock says, "everything is one. There is no way in which you draw a line between things. What we [normally] do is to make these subdivisions, but they're not real." Though McClintock arrived at this sense of oneness by focusing on parts (in particular on the chromosome) with an almost reductionist fervor, her approach is not reductionistic or "objective" in the traditional sense. "I found that the more I worked with them the bigger and bigger [the chromosomes] got, and when I was really working with them I wasn't outside, I was down there. I was part of the system." Like a Taoist sage, perhaps like a Taoist Yellow Emperor, McClintock's attitude is ironic: Both reductionist and holist, she strives to get to the bottom of things which she is aware have no bottom. In her sense of the whole, which she calls "a feeling for the organism," she revels in the uncertainties, interrelationships, and mutual dependencies that pervade nature. Her biographer describes her "access to the profound connectivity of all biological forms—of the cell, of the organism, of the ecosystem. The flip side of the coin is her conviction that, without an awareness of the oneness of things, science can give us only nature-in-pieces;

more often it gives us only pieces of nature. In McClintock's view, too restricted a reliance on scientific methodology invariably leads us into difficulty. 'We've been spoiling the environment just dreadfully and thinking we were fine, because we were using the techniques of science. Then it turns into technology, and it's slapping us back because we didn't think it through. We were making assumptions we had no right to make. From the point of view of how the whole thing actually worked, we know how part of it worked. . . . We didn't even inquire, didn't even see how the rest was going on. All these other things were happening and we didn't see it.' "

McClintock has evidently entered the turbulent mirror into a universe that is vaster, more complex, more fluid, less secure, and in a sense, more frightening than the one that has been portrayed by reductionist science. But in another sense, she seems to know that the turbulent universe is none of these things; it's a friendly place because we are all in it together.

FOREWORD

Why, if I ever did fall off . . . What will remain of me? . . . No one knows how this happened, and the Rig Veda speculates that possibly even the One does not know. . . . Alice interrupted, rather unwisely. . . . Not Chaos-like, together crushed and bruised, but, as the world harmoniously confused: Where order in variety we see, and where, though all things differ, all agree. . . . Alice could hardly help laughing. . . .

The Yellow Emperor got it and ascended to the cloudy heavens.

CHUANG TZU

BIBLIOGRAPHY

Following are some of the resources we used in writing this book. The list is meant to be more suggestive than exhaustive of the growing body of scientific and popular literature covering the topics discussed in *Turbulent Mirror*.

All epigraphs about the Yellow Emperor are from two sources: (1) A. C. Graham's translation of *The Book of Lieh-tzu*. London: John Murray, 1960; and (2) Burton Watson's translation of *The Complete Works of Chuang Tzu*. New York: Columbia University Press, 1968.

The Campbell epigraph is from his book with Bill Moyers, *The Power of Myth*. New York: Doubleday, 1988.

The Pope epigraph is from his "Essay on Man."

The Stevens epigraph is from his poem "Connoisseur of Chaos."

FOREWORD

Briggs, John, and F. David Peat. *Looking Glass Universe*. New York: Simon & Schuster, 1984.

Gleick, James. *Chaos: Making a New Science*. New York: Viking, 1987.

Kneale, Dennis. "Market Chaos: Scientists Seek Pattern in Stock Prices." *Wall Street Journal*, 17 Nov. 1987.

Prologue An Ancient Tension

Beir, Ulli, ed. *The Origin of Life and Death*. London: Heinemann, 1966.

Colum, Padraic. *Myths of the World*. New York: Grosset & Dunlap, 1959.

Disorder and Order: Proc. Stanford Int. Symp. (Sept. 14–16, 1981), Stanford Literature Studies 1. Saratoga, Calif.: Anima Libri, 1984.

Long, Charles H. *Alpha: The Myths of Creation*. New York: George Braziller, 1983.

Richardson, George P. "The Feedback Concept in American Social Science with Implications for Systems Dynamics." Int. Systems Dynamics Conf., July 1983.

Weiner, Philip, ed. *Dictionary of the History of Ideas* New York: Charles Scribner's Sons, 1973.

Chapter 1 Attractors and Reading Maps

Abraham, Ralph H., and Christopher D. Shaw. *Dynamics—The Geometry of Behavior, Part 1: Periodic Behavior*. Santa Cruz: Aerial Press, 1984.

———. *Dynamics—The Geometry of Behavior, Part 2: Chaotic Behavior*. Santa Cruz: Aerial Press, 1984.

Arnold, V. I. "Small Denominators and Problems of Stability of Motion in Classical and Celestial Mechanics." *Russ. Math. Surv.* 18 (1963): 85.

Goldstein, Herbert. *Classical Mechanics*. Reading, Mass.: Addison-Wesley, 1950.

Kolmogorov, A. N. "Preservation of Conditionally Periodic Movements with Small Change in the Hamiltonian Function." *Lecture Notes in Phys.* 93 (1979): 51.

Leung, A. "Limiting Behavior for a Prey-Predator Model with Diffusion and Crowding Effects." *Jour. Math. Biol.* 6 (1978): 87.

Smith, J. Maynard. *Models in Ecology*. Cambridge: Cambridge University Press, 1974.

———. The Theory of Games and the Evolution of Animal Conflicts." *Jour. Theor. Biol.* 47 (1974): 209.

Walker, Grayson H., and Joseph Ford. "Amplitude Instability and Ergodic Behavior for Conservative Nonlinear Systems." *Phys. Rev.* 188 (1969): 87.

Chapter 2 Turbulence, That Strange Attractor

Hammer, Signe, and Margaret L. Silbar. "The Riddle of Turbulence." *Science Digest*, May 1984.

Hénon, Michel. "A Two-dimensional Mapping with a Strange Attractor." *Commun. Math. Phys.* 130 (1976): 69.

Hopf, E. "A Mathematical Example Displaying Features of Turbulence." *Comm. Pure Appl. Math.* 1 (1948): 303.

Landau, L. D. *On the Problem of Turbulence: Collected Papers of* L. D. *Landau*. Trans. D. ter Haar. N.Y.: Pergamon, 1965.

Lorenz, Edward N. "Deterministic Nonperiodic Flow." *Jour. Atmospheric Sciences* 20 (1976): 69.

"The Mathematics of Mayhem." *The Economist*, 8 Sept. 1984.

Reiter, Carla. "The Turbulent Nature of a Chaotic World." *New Scientist*, 31 May 1984.

Ruelle, David, and Floris Takens. "On the Nature of Turbulence." *Commun. Math. Phys.* 20; (1971): 167.

Chapter 3 Doubling Route to(o) Strange

Feigenbaum, Mitchell J. "Quantitative Universality for a Class of Nonlinear Transformations." *Jour. Statistical Phys* 19 (1978): 25.

Hirsch, J. E., B. A. Huberman, and D. J. Scalapino. "Theory of Intermittency." *Phys. Rev.* 25 (1982): 519.

Jensen, Roderick V. "Classical Chaos." *American Scientist*, March–April 1987.

Markoff, John. "In Computer Behavior, Elements of Chaos." *New York Times*, 11 Sept. 1988.

May, Robert M. "Simple Mathematical Models with Very Complicated Dynamics." *Nature* 261 (1976): 459.

Saperstein, Alvin M. "Chaos—A Model for the Outbreak of War." *Nature* 309 (1984).

Screenivasan, K. R., and R. Ramshankar. "Transition Intermittency in Open Flows, and Intermittency Routes to Chaos." *Physica* 23D (1986): 246

Taubes, Gary. "Mathematics of Chaos." *Discover*, Sept. 1984.

Chapter 4 Iterative Magic

Chaitin, Gregory J. "Gödel's Theorem and Information." *Int. Jour. Theor. Phys.* 21, No. 12 (1982): 94.

Crutchfield, James P., J. Doyne Farmer, Norman H. Packard, and Robert Shaw. "Chaos." *Scientific American*, Dec. 1986.

Day, Richard H. "The Emergence of Chaos from Classical Economic Growth." *Quart. Jour. Econ.*, May 1983.

———. "Irregular Growth Cycles." *The American Econ. Rev.*, June 1982.

Hofstadter, Douglas R. *Gödel, Escher, Bach: An Eternal Golden Braid*. New York: Vintage Books, 1980.

CHAPTER 0 On Both Sides/Sides Both On

Batty, Michael. "Fractals—Geometry Between Dimensions." *New Scientist*, 4 April 1985.

———. *Microcomputer Graphics*. London: Chapman and Hall Computing, 1987.

Barcellos, Anthony. Interview with Benoit Mandelbrot. In *Mathematical People: Profiles and Interviews*. Ed. Donald J. Albers and G. L. Alexanderson. Boston: Birkhauser, 1985.

Gleick, James. "The Man Who Reshaped Geometry." *New York Times Magazine*, 8 Dec. 1985.

Dewdney, A. K. "Computer Recreations." *Scientific American*, Aug. 1985.

Kalikow, Daniel N. *David Brooks' Investigation of the Mandelbrot Set*. Monograph. Framingham, Mass.: Prime Computer, 1985.

La Brecque, Mort. "Fractal Symmetry." *Mosaic*, Jan/Feb. 1985.

Lorenz, Konrad. *On Agression*, trans., M. K. Wilson. New York: Harcourt, Brace & World, 1966.

Mandelbrot, Benoit. *The Fractal Geometry of Nature*. San Francisco; W. H. Freeman, 1982.

———. "The Many Faces of Scaling: Fractals, Geometry of Nature, and Economics." In *Self Organization and Dissipative Structures*. Eds. William C. Schieve and Peter M. Allen. Austin: University of Texas, 1982.

———. An interview. *Omni*, 5 Feb. 1984.

Mandelbrot, Benoit, Dann E. Passoja, and Alvin J. Paullay. "Fractal Character of Fracture Surfaces of Metals." *Nature* 308 (1984): 721.

Peterson, Ivars. "Packing It In." *Science News*, 2 May 1987.

Peitgen, H. O., and P. H. Richter. *The Beauty of Fractals*. Berlin: Springer-Verlag, 1986.

Poston, Tim, and Ian Stewart. *Catastrophe Theory and Its Applications*. Boston: Pitman, 1981.

Ruelle, David. "Strange Attractors." *Math Intell.* 2 (1980): 126.

Saunders, P. T. *An Introduction to Catastrophe Theory*. Cambridge: Cambridge University Press, 1980.

Shannon, C. E., and W. Weaver. *The Mathematical Theory of Information*. Urbana: University of Illinois Press, 1949.

Thom, René. *Structural Stability and Morphogenesis*. Trans., D. H. Fowler. Reading, Mass.: Benjamin, 1975.

Thompson, D'Arcy. *On Growth and Form*. Cambridge: Cambridge University Press, 1917.

Ullman, Montague. "Wholeness and Dreaming." In *Quantum Implications*. Eds., Basil Hiley and F. David Peat. London: Routledge and Kegan Paul, 1987.

Vilenkin, N. Ya. *Stories About Sets*. New York: Academic Press, 1965.

West, Bruce J., and Ary L. Goldberger. "Physiology in Fractal Dimensions." *American Scientist*, July/Aug. 1987.

Zeeman, E. C. *Catastrophe Theory*. Reading, Mass.: Addison-Wesley, 1977.

Chapter 4 The Great Wave

Bishop, A. R., and T. Schneider. *Solitons and Condensed Matter Physics*. New York: Springer-Verlag, 1978.

Dodd, R. K., J. C. Elibeck, J. D. Gibbon, and H. C. Morris. *Solitons and Nonlinear Wave Equations*. New York: Academic Press, 1982.

Fermi, Enrico, J. Pasta, and S. Ulam. "Studies of Nonlinear Problems." Reprinted in *Collected Papers of Enrico Fermi*. Vol. 2. Ed. E. Segre. Chicago: University of Chicago Press, 1965.

Frampton, Paul H. "Vacuum Instability and Higgs Scalar Mass." *Phys. Rev. Lett.* 37 (1976):1378.

———. "Consequences of Vacuum Instability in Quantum Field Theory." *Phys. Rev. Lett.* D15 (1977): 2922.

Ingersoll, Andrew P. "Models of Jovian Vortices." *Nature* 331 (1988):654.

Lee, T. D. *Particle Physics and Introduction to Field Theory*. Switzerland: Harwood Academic Pubs., 1981.

Lonngren, Karl, and Alwyn Scott, eds. *Solitons in Action*. New York: Academic Press, 1978.

Russell, J. Scott. "Report on Waves." *Report Brit. Assn. Advancement Sci.*, 1945.

Takeno, S., ed. *Dynamical Problems in Soliton Systems*. New York: Springer-Verlag, 1985.

Yuan, H. C., and B. M. Lake. "Nonlinear Deep Waves." In *The Significance of Nonlinearity in Natural Sciences*. Ed. B. Kursunoglu, A. Perlmutter, and L. F. Scott. New York: Plenum, 1977.

Chapter 3 Time's Arrow

Engel, Peter. "Against the Currents of Chaos." Rev. of *Order Out of Chaos*. *The Sciences*, Sept.–Oct. 1984.

Davies, Paul. *The Physics of Time Asymmetry*. London: Surrey University Press, 1974.

Landsberg, P. T. *Thermodynamics*. New York: Interscience, 1961.

Pagels, Heinz R. "Is the Irreversibility We See a Fundamental Property of Nature?" Rev. of *Order Out of Chaos*. *Phys. Rev.*, Jan. 1985.

Penrose, Oliver. "Improving on Newton." Rev. of *Order Out of Chaos*. *Nature*, 26 July 1984.

Prigogine, Ilya. *From Being to Becoming*. San Francisco: W. H. Freeman, 1980.

———. Interview with the authors. University of Texas, Austin, 27–29 April 1985.

Prigogine, Ilya, and Isabelle Stengers. *Order Out of Chaos*. Toronto: Bantam Books, 1984.

Prigogine, Ilya, and Y. Elskens. "From Instability to Irreversibility." *Proc. Natl. Academy of Sci.* 83 (1986): 5756.

Chapter 2 Feedback's Triumphs

Augros, Robert, and George Stanciu. *The New Biology: Discovering the Wisdom of Nature*. Boston: Shambhala, 1987.

Bergström, R. M. "An Analysis of the Information-carrying System of the Brain." *Synthese* 17 (1967): 425.

———. "An Entropy Model of the Developing Brain." *Developmental Psychobiology* 2 (3): 139.

———. "Quantitative Aspects of Neural Macrostates." *Cybernetic Medicine* (1973): 9.

Bertalanffy, Ludwig von. *General System Theory*. New York: George Braziller, 1968.

Büttner, Peter. "What Is Systems Thinking?" *The Brattleboro Bulletin*, July 1985.

Cairns, J., J. Overbaugh, and S. Miller. "The Origin of Mutants." *Nature* 335 (1988): 142.

Capra, Fritjof. *The Turning Point*. Toronto: Bantam, 1982.

Chandler, David L. "Rethinking Evolution." *Boston Globe*, 28 July 1986.

Edelman, Gerald. "Group Selection as the Basis for Higher Brain Function." In *The Organization of the Cerebral Cortex*. Eds. F. O. Schmitt et al. Cambridge: MIT Press, 1983.

Ellis, William. "Transnational Networks and World Order." *Transnational Perspectives* 8, No. 4 (1982): 9.

Fairfield, Roy. *Person-Centered Graduate Education*. Buffalo: Prometheus, 1977.

Fleischaker, Gail R. *Autopoiesis: System Logic of the Origin and Diversity of Life*. Doctoral thesis, Boston University, 1987.

Freeman, Walter J. "Physiological Basis of Mental Images." *Biol. Psychiatry* 18, No. 10 (1983): 1107.

Freeman, Walter J., and Christine A. Skarda. "Spatial EEG Patterns, Nonlinear Dynamics and Perception: The Neo-Sherrington View." *Brain Research Rev.* 10 (1985): 147.

Ferguson, Marilyn. *The Aquarian Conspiracy*. Los Angeles: Tarcher, 1980.

Forrester, Jay. *Urban Dynamics*. Cambridge: MIT Press, 1969.

———. "Common Foundations Underlying Engineering Management." The Institute of Electrical and Electronics Engineers. Reprint of talk.

Fox, Jeffrey L. "The Brain's Dynamic Way of Keeping in Touch." *Science*, Aug. 1984.

Gardner, Howard. *The Mind's New Science*. New York: Basic Books, 1985.

Garfinkel, Alan. "A Mathematics for Physiology." *American Physiological Society* (1983).

Gray, William. "Understanding Creative Thought Process: An Early Formulation of the Emotional-Cognitive Structure Theory." *Man-Environment Systems* 9, No. 1 (1979).

Globus, Gorden G. "Three Holonomic Approaches to the Brain." In *Quantum Implications*. Eds. Basil Hiley and F. David Peat. London: Routledge and Kegan Paul, 1987.

Gould, Stephen Jay, "Kropotkin was No Crackpot." *Natural History* (1988): 12–21.

Harley, Richard. "Global Networks: Trading Recipes and Technologies from Maine to Nepal." *Christian Science Monitor*, 7 Oct. 1982.

Henderson, Hazel. *The Politics of the Solar Age*. Garden City, N.Y.: Doubleday, 1981.

Hofstadter, Douglas R. *Metamagical Themas: Questing for the Essence of Mind and Pattern*. Toronto: Bantam, 1985.

Hooper, Judith, and Dick Teresi. *The 3-Pound Universe*. New York: Macmillan, 1986.

Jantsch, Erich. *The Self-Organizing Universe*. Oxford: Pergamon, 1980.

Johnson, George. "Learning, Then Talking." *New York Times*. 16 Aug. 1988.

Joseph, Laurence E. "Britain's Whole Earth Guru." *New York Times Magazine*, 2 Nov. 1986.

Keller, Evelyn Fox. *A Feeling for the Organism*. New York: W. H. Freeman and Co., 1983.

King, Roy, Joachim D. Raese, and Jack Barchas. "Catastrophe Theory of Dopaminergic Transmission: A Revised Dopamine Hypothesis of Schizophrenia." *Jour. Theoretical Biol.* 92 (1981): 373.

LaViolette, Paul A. "Thoughts about Thoughts about Thoughts: The Emotional Perceptive Cycle Theory." *Man-Environment Systems* 9, No. 1 (1979).

Lipnack, Jessica, and Jeffrey Stamps. *Networking: The First Report and Directory*. Garden City, N.Y.: Doubleday, 1982.

Lovelock, J. E. *Gaia: A New Look at Life on Earth*. Oxford: Oxford University Press, 1979.

Margulis, Lynn, and Dorion Sagan. *Microcosmos*. New York: Summit Books, 1986.

Meadows, Donella H., et al. *The Limits to Growth*. New York: New American Library, 1974.

———. "Whole Earth Models and Systems." *The Co-Evolution Quarterly Report*. Reprint.

Miskin, Mortimer, and Tim Appenzeller. "The Anatomy of Memory." *Scientific American*, June 1987.

Mosekilde, Erik, Javier Aracil, and Peter M. Allen. "Chaotic Behavior in Non-linear Dynamic Models." Manuscript.

Naisbitt, John. *Megatrends*. New York: Warner, 1984.

Pagels, Heinz. *The Dreams of Reason*. New York: Simon & Schuster, 1988.

Peat, F. David. *Artificial Intelligence*. New York: Baen, 1988.

Peters, Thomas, and Robert H. Waterman, Jr. *In Search of Excellence*. New York: Harper & Row, 1982.

———. *Thriving on Chaos: A Handbook for a Managerial Revolution*. New York: Alfred A. Knopf, 1987.

Rosenfield, Israel. "Neural Darwinism: A New Approach to Memory and Perception." *New York Rev. of Books*, 9 Oct. 1986.

Senge, Peter M. "Systems Dynamics, Mental Models and the Development of Management Intuition." Int. System Dynamics Conf., July 1985.

———. "Systems Thinking in Business: An Interview with Peter Senge." *Revision* 7, No. 2.

———. Interviews with J. Briggs. MIT, Cambridge, Mass., 4 Sept. 1986 and 8 Jan. 1987.

Sterman, John D. "Deterministic Chaos in Models of Human Behavior: Methodological Issues and Experimental Results." Manuscript.

Tank, David W., and John J. Hopfield. "Collective Computation in Neuronlike Circuits." *Scientific American*, Dec. 1987.

Theobald, Robert. *The Rapids of Change*. Indianapolis: Knowledge Systems, 1987.

Varela, Francisco J. *Principles of Biological Autonomy*. New York: North Holland, 1979.

Weiner, Jonathan. "In Gaia's Garden." *The Sciences* (Jan.–Feb. 1986): 2.

Weisburd, Stefi. "Neural Nets Catch the ABCs of DNA." *Science News*, 1 Aug. 1987.

Young, J. Z. "Hunting the Homunculus." *New York Rev. of Books*, 4 Feb. 1988.

Chapter 1 Quantum Roots to Strange

Bohm, David, B. Hiley, and P. N. Kaloyerou. "An Ontological Basis for Quantum Theory." *Phys. Reports*. 144, No. 6 (1987): 323.

D'Espagnat, Barnard. "Quantum Theory and Reality." *Scientific American*, Nov. 1979.

Peat, F. David. "The Emergence of Structure and Organization from Physical Systems." *Int. Jour. Quantum Chemistry*, Quantum Biology Symposium No. 1 (1974): 213.

———. "The Evolution of Structure and Order in Quantum Mechanical Systems." *Collective Phenomena* 2 (1976): 149.

———. "Time, Structure and Objectivity in Quantum Theory." *Foundations of Physics*, December, 1988.

———. *Superstrings and the Search for the Theory of Everything*. New York: Contemporary Books, 1988.

Shimony, Abner. "The Reality of the Quantum World." *Scientific American*, Jan. 1988.

Wheeler, John A. "Bits, Quanta and Meaning." Preprint.

Wheeler, John A., and Wojciech H. Zurek, eds. *Quantum Theory and Measurement*. Princeton: Princeton University Press, 1983.

Prologue Tension Forever New

Arnheim, Rudolf. *Entropy and Art: An Essay on Disorder and Order*. Berkeley: University of California Press, 1971.

Bernstein, Leonard. *The Unanswered Question*. Cambridge, Mass.: Harvard University Press, 1976.

Briggs, John. *Fire in the Crucible: The Alchemy of Creative Genius*. New York: St. Martin's Press, 1988.

———. "Reflectaphors: The (Implicate) Universe as a Work of Art." In *Quantum Implications*. Eds. Basil Hiley and F. David Peat. London: Routledge and Kegan Paul, 1987.

Bohm, David and F. David Peat. *Science, Order and Creativity*. New York: Bantam, 1987.

Dodge, Charles and Curtis R. Bahn. "Musical Fractals." *Byte*, June 1986.

Gruber, Howard. "Inching Our Way Up Mount Olympus: The Evolving Systems Approach to Creative Thinking." In *The Nature of Creativity*. Ed. Robert J. Sternberg. Cambridge: Cambridge University Press, 1988.

Hadamard, Jacques. *The Psychology of Invention in the Mathematical Field*. New York: Dover, 1945.

Holton, Gerald. *Thematic Origins of Scientific Thought: Kepler to Einstein*. Cambridge, Mass.: Harvard University Press, 1973.

James, Henry. *The Art of the Novel*. New York: Charles Scribner's Sons, 1934.

Koestler, Arthur. *The Act of Creation*. New York: Macmillan, 1964.

Monaco, Richard, and John Briggs. *The Logic of Poetry*. New York: McGraw-Hill, 1974.

Paz, Octavio. *The Monkey Grammarian*. Trans. Helen R. Lane. New York: Grove Press, 1981.

Schoenberg, Arnold. *Style and Idea*. New York: Philosophical Library, 1950.

Thomsen, Dietrick E. "Making Music—Fractally." *Science News*, 22 March 1980.

FOREWORD

Keller, Evelyn Fox. *A Feeling for the Organism*. New York: W. H. Freeman and Co., 1983.

Rifkin, Jeremy. *Algeny*, New York: Viking, 1983.

ILLUSTRATION CREDITS

All the chapter headings were designed by Cindy Tavernise.

FRONTISPIECE

Cindy Tavernise

Prologue

P.1	Details from Ch'en Jung, *Nine Dragons*; Francis Gardner Curtis Fund; courtesy Museum of Fine Arts, Boston
P.2	Courtesy Norman J. Zabusky, reprinted from *Physics Today*, © 1984
P.3	Cindy Tavernise
P.4	Cindy Tavernise

Chapter 1

1.1	Cindy Tavernise
1.2	Cindy Tavernise
1.3	Cindy Tavernise
1.4	Cindy Tavernise
1.5	Cindy Tavernise
1.6	Cindy Tavernise
1.7	Cindy Tavernise
1.8	Cindy Tavernise
1.9	Cindy Tavernise
1.10	Cindy Tavernise
1.11	Cindy Tavernise
1.12	Cindy Tavernise
1.13	Cindy Tavernise
1.14	Cindy Tavernise
1.15	Cindy Tavernise
1.16	Cindy Tavernise
1.17	Cindy Tavernise
1.18	Cindy Tavernise
1.19	Cindy Tavernise
1.20	Cindy Tavernise
1.21	Courtesy Jet Propulsion Laboratory, Pasadena, California

Chapter 2

2.1	Cindy Tavernise
2.2	Windsor Castle, Royal Library, © Her Majesty Queen Elizabeth II
2.3	Cindy Tavernise
2.4	Cindy Tavernise
2.5	Cindy Tavernise
2.5	Cindy Tavernise
2.6	Cindy Tavernise
2.7	Cindy Tavernise
2.8	Cindy Tavernise
2.9	Cindy Tavernise

Chapter 3

3.1	Cindy Tavernise
3.2	Cindy Tavernise
3.3	Cindy Tavernise
3.4	Cindy Tavernise
3.5	Cindy Tavernise
3.6	Cindy Tavernise
3.7	Cindy Tavernise
3.8	Leo Kadanoff, from *Physics Today*, © 1983
3.9	Courtesy Roderick V. Jensen, reprinted from *American Scientist*, © 1987; embellishing sketches by Cindy Tavernise
3.10	Courtesy Roderick V. Jensen, reprinted from *American Scientist*, © 1987

Chapter 4

4.1	Cindy Tavernise
4.2	Cindy Tavernise
4.3	James Crutchfield
4.4	Douglas Smith, Boston Museum of Science

4.5 Cindy Tavernise, after plots made by James Crutchfield
4.6 Cindy Tavernise

Double-Page Alice

Cindy Tavernise

CHAPTER 0

0.1 Cindy Tavernise
0.2 Cindy Tavernise
0.3 Cindy Tavernise
0.4 Courtesy Roderick V. Jensen, reprinted from *American Scientist*, © 1987
0.5 Courtesy Benoit Mandelbrot, IBM Research
0.6 Courtesy Benoit Mandelbrot, IBM Research
0.7 Cindy Tavernise
0.8–0.18 These illustrations were computed on a Prime 9955 semiminicomputer and imaged in 300 dpi halftone via PostScript on an Apple LaserWriter printer. A Mandelbrot Set computation was made for each of the 90,000 dots per square inch of these pictures. They are reproduced here with the kind permission of Prime Computer Inc., Natick, Mass., and of the developers of this process at Prime: David Brooks and Dan Kalikow. Imaging technique, © Prime Computer Inc.
0.19 Cindy Tavernise
0.20 Cindy Tavernise using plots made by Roderick V. Jensen, reprinted from *American Scientist*, © 1987, and Leo Kadanoff from *Physics Today*, © 1983
0.21 Courtesy H-O. Peitgen and P. H. Richter, reprinted from *The Beauty of Fractals*, Springer-Verlag, Heidelberg, © 1986
0.22 Cindy Tavernise
0.23 Douglas Smith, Boston Museum of Science
0.24 Courtesy of Michael Batty, reprinted from his *Microcomputer Graphics*, Chapman and Hall Computing, and *New Scientist*, © 1985
0.25 Courtesy of Michael Batty, reprinted from his *Microcomputer Graphics*, Chapman and Hall Computing, and *New Scientist*, © 1985
0.26 Reprinted by permission of The British Museum
0.27 Katsushika Hokusai, *The Great Wave*, Spaulding Collection; courtesy Museum of Fine Arts, Boston

Reverse Double-Page Alice

Cindy Tavernise

Chapter 4

4.1 Cindy Tavernise
4.2 Cindy Tavernise
4.3 Cindy Tavernise
4.4 Jet Propulsion Laboratory, Pasadena, California
4.5 Cindy Tavernise

4.6	Cindy Tavernise
4.7	Cindy Tavernise
4.8	NASA

Chapter 3

3.1A	John Briggs and F. David Peat, reprinted from *Looking Glass Universe*
3.1B, C	We thank Dr. Ralph Abraham for the illustrations of Bénard flow. They are taken from his publication, *Dynamics: The Geometry of Behavior: Chaotic Behavior*, published by Aerial Press, Inc.
3.2	Cindy Tavernise
3.3	Cindy Tavernise
3.4	Photographs by Friz Goro from *Being to Becoming* by Ilya Prigogine, © 1980 W. H. Freeman and Company; reprinted with permission
3.5	Courtesy National Optical Astronomy Observatories
3.6	John Briggs
3.7	Cindy Tavernise

Chapter 2

2.1	Cindy Tavernise, after Lynn Margulis
2.2	Cindy Tavernise, after John Lovelock
2.3	Cindy Tavernise, after Erich Jantsch
2.4	Cindy Tavernise, after Matti Bergström
2.5	Cindy Tavernise, after Michael Merzenich
2.6	Cindy Tavernise, after Peter Senge

Chapter 1

1.1	Cindy Tavernise

Prologue

P.1	James Crutchfield
P.2	Reprinted by permission Sterling Lord Literistic, Inc., © 1964 by Arthur Koestler
P.3	Reprinted by permission Sterling Lord Literistic, Inc., © 1964 by Arthur Koestler
P.4	Cindy Tavernise
P.5	Metropolitan Museum of Art, New York, Bequest of Benjamin Altman, 1913. All rights reserved, The Metropolitan Museum of Art

Endispiece

Cindy Tavernise

INDEX